Data Usability in the Enterprise

How Usability Leads to Optimal Digital Experiences

Praveen Gujar

Apress®

Data Usability in the Enterprise: How Usability Leads to Optimal Digital Experiences

Praveen Gujar
Independent Researcher
Saratoga, CA, USA

ISBN-13 (pbk): 979-8-8688-1182-1   ISBN-13 (electronic): 979-8-8688-1183-8
https://doi.org/10.1007/979-8-8688-1183-8

Managing Director, Apress Media LLC: Welmoed Spahr
Acquisitions Editor: Aditee Mirashi
Development Editor: James Markham
Editorial Assistant: Kripa Joseph
Copy Editor: Kimberly Burton-Weisman

Cover designed by eStudioCalamar

Distributed to the book trade worldwide by Springer Science+Business Media New York, 1 New York Plaza, Suite 4600, New York, NY 10004-1562, USA. Phone 1-800-SPRINGER, fax (201) 348-4505, e-mail orders-ny@springer-sbm.com, or visit www.springeronline.com. Apress Media, LLC is a California LLC and the sole member (owner) is Springer Science + Business Media Finance Inc (SSBM Finance Inc). SSBM Finance Inc is a **Delaware** corporation.

For information on translations, please e-mail booktranslations@springernature.com; for reprint, paperback, or audio rights, please e-mail bookpermissions@springernature.com.

Apress titles may be purchased in bulk for academic, corporate, or promotional use. eBook versions and licenses are also available for most titles. For more information, reference our Print and eBook Bulk Sales web page at http://www.apress.com/bulk-sales.

Any source code or other supplementary material referenced by the author in this book is available to readers on GitHub. For more detailed information, please visit https://www.apress.com/gp/services/source-code.

If disposing of this product, please recycle the paper

*To my sons: May your curiosity never fade,
and your thirst for knowledge never be quenched.*

*To my wife and parents: My endless gratitude for the love
and support that makes all things possible.*

—Praveen

Table of Contents

About the Author

Praveen Gujar stands at the forefront of product innovation, boasting an illustrious career of close to two decades and marked by the successful launch of cutting-edge enterprise data products in digital advertising. A stalwart in the tech industry, Praveen has left his mark on tech giants like LinkedIn, Amazon, and Twitter, spearheading initiatives that have catapulted these companies into multi-billion-dollar echelons.

A globally recognized thought leader, Praveen shares insights at prominent industry and academic conferences, expert panels, and publications in popular platforms such as Forbes and IEEE, shaping industry trends and guiding senior executives in AI, digital transformation, and product innovation. He holds fellowships with esteemed organizations such as Engineering and Technology (IET) and The Institution of Electronics and Telecommunication Engineers (IETE) and serves as marketing chair for IEEE Technology and Engineering Management Society (TEMS), further cementing his influence in the tech community. Praveen's work inspires innovation and strategic growth across the tech landscape.

About the Technical Reviewer

 **Sanyam Jain** is a distinguished Cloud Security Engineer with extensive expertise in cybersecurity. Known for his dedication to securing digital environments, Sanyam's achievements reflect his deep commitment to protecting critical infrastructure and contributing to the broader security community. He has excelled across cloud security, security operations, application security, compliance, and security automation, consistently developing strategies that help organizations exceed their security objectives.

His technical proficiency spans network security, threat detection, data encryption, and access control, with expertise across AWS, Azure, and Google Cloud. Sanyam's discovery of security vulnerabilities has earned recognition in leading publications like *Forbes*, *TechCrunch*, *ZDNet*, and *Bleeping Computer*, underscoring his thought leadership in the field.

In addition to his corporate work, Sanyam collaborates with NGOs such as GDI Foundation and CSIRT.Global, supporting the protection of critical Internet infrastructures. He also serves as a judge for Bintelligence and has evaluated projects in India's largest hackathon. Academically, he holds a master's degree in technology from BITS Pilani, graduating with distinction, and is CKA certified in Kubernetes. His contributions extend to book reviews on cybersecurity and cloud computing, reflecting his commitment to advancing knowledge and best practices in the industry.

Acknowledgments

This book would not have been possible without the dedication and expertise of its incredible contributing authors. A heartfelt "thank you" to Sriram Panyam, Anujkumarsinh Donvir, Fardin Quazi, Pradeep Chintale, Sivanagaraju Gadiparthi, Gunjan Pariwal, and Chhaya Kewalramani for sharing their knowledge and insights. Their contributions have enriched this work immeasurably.

Introduction

This book explores the critical principles of software usability, data quality, and security—integral aspects of designing effective, user-centered digital systems. With an overarching focus on the interplay between usability, accessibility, and data governance, this book is a comprehensive guide for software designers, data managers, and decision-makers in today's data-centric landscape.

Purpose and Audience

This book is crafted for professionals and students in software development, data management, and IT governance, aiming to deepen their understanding of how to design software and manage data systems that prioritize both functionality and user satisfaction. Its insights are also valuable for business leaders and data stewards responsible for maintaining data quality and compliance while delivering seamless user experiences. Whether you are a software developer, UX designer, or data governance expert, this book equips you with the foundational concepts and practical tools to balance the demands of security, accessibility, and usability.

Structure of the Book

The book is organized into ten chapters, each focusing on a unique aspect of creating user-friendly, reliable, and secure software and data systems:

INTRODUCTION

Chapter 1 introduces the foundational principles of software usability, explaining why it is a crucial determinant of software success. It defines usability in terms of ease of learning, user satisfaction, and its influence on productivity, setting the stage for subsequent chapters that dive deeper into design principles and data strategies.

Chapter 2 explores core design principles such as clarity, modularity, reusability, and iterative design. It highlights how these principles can help create robust, maintainable, and scalable software systems, ultimately leading to better user experiences and streamlined development processes.

Chapter 3 emphasizes inclusive design, advocating for software that serves users of diverse abilities and needs. This chapter introduces accessibility standards like WCAG and explores how accessibility enhances usability across various contexts, including education, healthcare, and e-commerce.

Chapter 4 focuses on the importance of standardized data formats and interoperability between systems. These concepts enable seamless data exchange, which is essential for collaborative work and informed decision-making across industries.

Chapter 5 introduces data warehousing, integration, and visualization tools that improve data accessibility and usability. By using these tools, organizations can better structure, analyze, and present data in ways that support decision-making.

Chapter 6 delves into visualization strategies, illustrating how data can be represented visually to improve comprehension and facilitate actionable insights for end-users.

Chapter 7 outlines principles of user-centered design (UCD) to ensure software meets user needs. The chapter discusses methods like personas, user stories, and usability audits, which help teams maintain a user-first approach from development through deployment.

Chapter 8 addresses critical data quality components—accuracy, consistency, completeness, and reliability. It also covers common issues, like data decay and system synchronization problems, that can degrade data integrity.

Chapter 9 examines the role of governance frameworks in maintaining data quality, security, and usability. This chapter explores frameworks like DAMA-DMBOK and ISO/IEC 38500, which provide best practices for structuring data governance programs.

Chapter 10 explores the balance between data usability and privacy. It emphasizes techniques like differential privacy and federated learning to protect data while ensuring it remains accessible for analysis.

Chapter 11 introduces success stories from companies like Apple, Google, and Netflix, offering concrete examples of how usability can drive market success. These examples illustrate the impact of usability in fields like mobile technology, streaming, e-commerce, and productivity, highlighting lessons that can inspire similar results across various sectors.

Chapter 12 explores how emerging technologies like AI, AR/VR, and voice interfaces are shaping the future of usability. It delves into evolving design approaches that accommodate personalized, context-aware, and adaptive user experiences. The discussion anticipates the opportunities and challenges these advancements present, particularly in balancing usability and ethical considerations.

Chapter 13 emphasizes the role of change management in software usability, stressing that successful updates require user support and acceptance. This chapter details how aligning usability with change management practices—such as user training, communication, and feedback integration—can drive smoother transitions and improve overall user satisfaction.

Chapter 14 addresses common obstacles like cognitive overload, balancing functionality with simplicity, and accommodating diverse user needs. Outlining strategies to manage these challenges prepares readers to tackle the practical difficulties they may encounter when designing software for varied audiences.

Chapter 15 considers how usability principles will continue to evolve in response to technological advances and user expectations. It advocates for adaptive, human-centered approaches to usability, particularly in emerging areas like multimodal interfaces and AI-driven personalization.

Chapter 16 synthesizes the book's content, reaffirming the critical role of usability in modern software design and summarizing key takeaways for the future. It underscores the importance of maintaining core usability principles while adapting to new contexts and technologies. As usability becomes central to software success, this chapter encourages readers to view usability as both a technical and ethical commitment to users.

Key Themes

Throughout the book, themes of accessibility, adaptability, and ethical responsibility underscore the practical guidance offered. Readers will gain insights on balancing user needs with technological potential, understanding the value of usability beyond functionality, and implementing change management practices that sustain usability through system updates.

In sum, this book is an essential guide for creating software that is not only user-friendly but also future-ready. It offers a comprehensive framework to understand and apply usability principles that aligns with evolving technologies and diverse user needs, equipping readers to build systems that serve as powerful, intuitive tools in an increasingly digital world.

PART I

Introduction to Data Usability

CHAPTER 1

Concepts and Importance

In the rapidly evolving world of software development, one of the key determinants of a product's success is its usability. Whether it's a mobile application, enterprise software, or a web-based tool, usability defines how effectively and efficiently users can interact with a software system. Usability is not just about aesthetics or functionality; it is a comprehensive measure that encompasses ease of learning, user satisfaction, and the system's ability to meet user needs. This chapter delves deep into software usability, its importance, and its broader impact on software success.

The increasing complexity of software products, coupled with the growing demand for intuitive and seamless user experiences, has placed usability at the forefront of software design and development. Poor usability can lead to frustrated users, high abandonment rates, and a failure to meet business goals. Conversely, software with high usability leads to increased user satisfaction, higher productivity, and a competitive edge in the market.

Understanding software usability requires an exploration of its key principles, components, and best practices. This chapter introduces the fundamental concepts of usability, exploring why it is so critical in modern software design.

P. Gujar, *Data Usability in the Enterprise*, https://doi.org/10.1007/979-8-8688-1183-8_1

What Is Usability?

Usability, in the context of software design, refers to the ease with which users can interact with a product or system to achieve their goals efficiently, effectively, and satisfactorily. Usability encompasses various factors, including learnability, efficiency, memorability, error tolerance, and user satisfaction.

Key Definitions of Usability

- **Ease of use**: How easily new or experienced users can accomplish tasks using the software.

- **Effectiveness**: The accuracy and completeness with which users achieve their intended goals.

- **Efficiency**: The speed and effort required to complete a task in the software system.

- **Satisfaction**: The user's overall experience and comfort when interacting with the system.

Usability is often viewed as a subset of user experience (UX), focusing on the specific interactions between the user and the interface. It is about making software functional and ensuring that users can interact with it comfortably and intuitively.

In practice, usability involves understanding users' needs, preferences, and limitations and designing software that accommodates those factors. A system with poor usability may be functional but cumbersome, resulting in reduced productivity and frustration. In contrast, highly usable software supports users in achieving their goals smoothly and without unnecessary cognitive effort.

Key Components of Usability

Learnability

Learnability refers to how quickly new users can become proficient with the software. A system with high learnability allows users to easily understand its features and interface, reducing the time required to perform tasks. Intuitive navigation, clear instructions, and well-organized content all contribute to improving learnability.

Efficiency

Once users have learned the software, efficiency determines how quickly they can complete tasks. Efficient software minimizes unnecessary steps, reduces waiting times, and provides shortcuts or advanced options for experienced users. Efficiency is especially critical in professional or enterprise software, where the goal is often to maximize productivity.

Memorability

Memorability addresses how easily users can recall how to use the software after a period of not using it. Software with good memorability reduces the need for users to relearn processes, which is essential in applications that are not used frequently. Consistent design, predictable behaviors, and clear visual cues contribute to better memorability.

Error Tolerance

Even the best-designed systems can lead to user errors. Error tolerance refers to how well the software helps users recover from mistakes and how effectively it prevents errors from happening in the first place. Clear error messages, undo options, and the prevention of incorrect inputs can all contribute to improving error tolerance.

Satisfaction

User satisfaction is a subjective measure of how pleasant and satisfying the overall experience of using the software is. Factors such as visual design, user feedback, responsiveness, and the emotional appeal of the interface can influence satisfaction. Satisfaction is a critical component of long-term user engagement and loyalty.

Principles of Usability

Several guiding principles form the foundation of usable software design. These principles ensure that software is not only functional but also intuitive and user-friendly.

Consistency

Consistency across the software ensures that users can predict how different parts of the system will behave. For example, buttons should look and behave similarly throughout the application, and navigation should follow a predictable pattern. Consistency reduces the cognitive load on users and helps them learn the system more quickly.

Feedback

Users need feedback to understand the outcomes of their actions. Feedback can be in the form of visual changes (e.g., highlighting a button when clicked), auditory signals, or textual messages. Clear and timely feedback ensures that users know whether their actions were successful, preventing frustration.

Visibility of System Status

Users should always be aware of the current state of the system. For example, when a file is being uploaded, a progress bar informs users that the action is in progress. Ensuring that users know what is happening within the system at any given moment helps build trust and prevents confusion.

Affordances

Affordances are design elements that suggest how they should be used. For example, a button should look clickable, and a slider should suggest it can be moved. Clear affordances make software more intuitive and reduce the need for instructions.

Flexibility and Efficiency of Use

Good software design accommodates both novice and experienced users. While beginners may need guidance and simple interfaces, experienced users often appreciate shortcuts, customizable settings, and advanced features that allow them to perform tasks more quickly.

Error Prevention

A key aspect of usability is minimizing the possibility of user errors. This can be achieved by providing clear instructions, validating inputs before processing, and designing workflows that reduce the likelihood of mistakes. Preventing errors is often more effective than helping users recover from them.

Recovery from Errors

When errors do occur, the system should offer users clear guidance on how to recover. Providing undo options, clear error messages that explain what went wrong and how to fix it, and preventing catastrophic errors can greatly improve the user experience.

Benefits of High Usability in Software

High usability brings several tangible and intangible benefits, making it an essential focus for software developers, businesses, and users alike.

Increased User Satisfaction

Software that is easy to use and meets user needs results in higher levels of satisfaction. Satisfied users are more likely to continue using the product, recommend it to others, and provide positive reviews, which in turn can increase the software's market share and reputation.

Improved Productivity

When users can accomplish their tasks more quickly and with less effort, their productivity increases. This is particularly important for business applications, where efficiency can directly impact profitability. Well-designed software minimizes downtime, reduces errors, and allows users to focus on their core tasks rather than struggling with the interface.

Reduced Training and Support Costs

Highly usable software reduces the need for extensive training and support. If users can quickly learn how to use the software and encounter fewer issues, organizations can save time and money on training programs and technical support services. This is especially important in enterprise settings, where large teams may need to use the software.

Higher Adoption Rates

Software with poor usability is often abandoned by users, even if it has powerful features. High usability increases the likelihood of adoption, as users are more willing to engage with software that they find easy and enjoyable to use. This is particularly critical in competitive markets, where user-friendly software often has a significant advantage.

Lower Development and Maintenance Costs

Addressing usability issues during the design phase can reduce long-term development and maintenance costs. If usability is neglected early on, it can lead to complex, hard-to-fix issues later, requiring extensive rework. Designing for usability from the outset can prevent costly redesigns and patchwork fixes.

Usability vs. User Experience

While usability and user experience (UX) are closely related, they are distinct concepts in software design. Understanding the difference is crucial for developers and designers aiming to create a holistic, user-centered product.

Usability: A Subset of UX

Usability is a critical component of UX but not the entirety of it. Usability focuses on how easily users interact with the software, covering aspects such as efficiency, learnability, and error tolerance. In contrast, UX encompasses the broader emotional and psychological experience of using the software, including usability, aesthetics, accessibility, and overall satisfaction.

UX: The Bigger Picture

UX includes factors such as how the software looks, feels, and aligns with user goals and emotions. For example, an application may be highly usable in terms of task efficiency but still fail to create a positive UX if its design is visually unappealing or it doesn't align with users' emotional needs. UX designers take a more holistic approach, ensuring that the software not only works well but also provides an enjoyable and engaging experience.

Interplay Between Usability and UX

While usability is critical to achieving a good user experience, a well-rounded UX approach ensures that the software meets user needs on a deeper level. Usability and UX must work together to create software that is functional and efficient as well as satisfying and meaningful to the user.

Usability Heuristics

Usability heuristics are general principles or best practices that guide the design of user-friendly systems. They serve as a foundation for evaluating and improving usability in software systems.

Jakob Nielsen's 10 Usability Heuristics

One of the most widely referenced sets of usability heuristics was developed by Jakob Nielsen, a pioneer in usability research. These heuristics apply to various software systems and provide guidelines for enhancing usability.

- **Visibility of system status**: Keep users informed about what is happening in the system through appropriate feedback.

- **Match between the system and the real world**: Use familiar language and concepts that align with the users' real-world experiences.

- **User control and freedom**: Provide users with options to easily undo or exit unwanted actions.

- **Consistency and standards**: Ensure that design elements follow established conventions so users do not have to wonder what different controls mean.

- **Error prevention**: Prevent errors from occurring by offering guidance and validating inputs before processing.

- **Recognition rather than recall**: Reduce the cognitive load by making information and options easily accessible and visible.

- **Flexibility and efficiency of use**: Allow users to tailor their interactions based on experience levels, providing shortcuts for expert users.

- **Aesthetic and minimalist design**: Avoid overloading the interface with unnecessary information or functionality.

- **Help users recognize, diagnose, and recover from errors**: Present error messages that are clear, informative, and suggest solutions.

- **Help and documentation**: Provide easily accessible help or documentation to assist users in understanding how to use the software.

These heuristics serve as a benchmark for usability and can guide the design process to ensure that systems are intuitive, user-friendly, and efficient.

Usability Testing

Usability testing is a process used to evaluate how easily users can interact with a software system. This testing involves observing real users as they attempt to complete tasks using the software, allowing designers to identify usability issues and gather feedback.

Goals of Usability Testing

The primary goal of usability testing is to uncover problems in the user interface and overall user experience before the software is released to a wider audience. Usability testing helps answer questions like the following.

- Can users complete tasks easily and efficiently?

- Where do users encounter difficulties or confusion?

- How satisfied are users with the software?

- What improvements can be made to enhance usability?

Types of Usability Testing

There are several approaches to usability testing, each suited to different stages of the development process.

- **Moderated vs. unmoderated testing**: Moderated testing involves direct interaction between the user and a facilitator who guides the session, while unmoderated testing allows users to complete tasks independently.

- **Remote vs. in-person testing**: Remote testing allows users to test software in their natural environment, often providing more realistic insights. In-person testing, however, allows for direct observation and more immediate feedback.

- **Formative vs. summative testing**: Formative testing is conducted during the development process to refine the design, while summative testing is conducted at the end to assess overall usability.

Benefits of Usability Testing

Usability testing provides valuable insights into how real users interact with software. It allows developers to identify issues that may not be obvious through automated testing or developer reviews. The iterative nature of usability testing—testing, refining, and testing again—ensures that the final product is more user-friendly and effective.

Methods for Improving Usability

Improving software usability requires a combination of research, design principles, and iterative testing. Several methods can be employed to ensure that a software product is easy to use and meets user needs.

User Research

Understanding the target audience is crucial to designing usable software. Techniques such as user interviews, surveys, and ethnographic studies can provide insights into user behaviors, needs, and pain points. User personas and scenarios can be developed from this research to guide design decisions.

Wireframing and Prototyping

Before developing the final product, designers create wireframes and prototypes to visualize and test different design ideas. Wireframes are low-fidelity, simplified representations of the user interface, while prototypes are interactive mock-ups that allow users to experience the flow of the system. These tools help identify usability issues early in the design process.

Iterative Design

Usability improvements are best achieved through an iterative process. After each round of testing, designers refine the interface based on user feedback and retest the system to validate the changes. This continuous cycle of designing, testing, and refining ensures that usability issues are addressed before the software is fully developed.

Accessibility Considerations

Ensuring that software is accessible to all users, including those with disabilities, is a critical aspect of usability. Accessibility features such as keyboard navigation, screen reader compatibility, and customizable text sizes improve usability for a broader audience. Incorporating accessibility standards (such as the Web Content Accessibility Guidelines (WCAG)) into the design process ensures that the software is inclusive.

14

Role of Feedback in Usability

Feedback is essential for improving software usability. Gathering input from real users throughout the design and development process helps designers identify issues, validate design decisions, and ensure that the software meets user needs.

Types of User Feedback

- **Direct feedback**: Feedback from usability testing, interviews, or focus groups where users provide their thoughts and reactions to the system.

- **Indirect feedback**: Metrics and analytics gathered from user interactions, such as click patterns, time on task, or error rates, can reveal usability issues.

- **Post-launch feedback**: After the software is released, gathering feedback through surveys, customer support interactions, and app store reviews can highlight areas for improvement.

Incorporating Feedback into the Design Process

User feedback should be continuously incorporated into the design process, ensuring that changes and improvements are made based on real-world usage. Agile and iterative development processes are well-suited to incorporating feedback, allowing continuous refinement and adjustment throughout the project lifecycle.

Accessibility and Inclusive Design

Usability is closely related to accessibility and inclusive design, which aim to make software usable for people with varying abilities and needs. Designing for accessibility ensures that individuals with disabilities, such as visual impairments or motor difficulties, can use the software effectively.

Key Accessibility Features

- **Keyboard navigation**: Ensures that users can navigate the software without a mouse, using keyboard shortcuts or tabbing.

- **Screen reader support**: Provides alternative text for images, buttons, and other visual elements so that screen readers can convey this information to visually impaired users.

- **Adjustable text size and contrast**: Allows users to customize the text size or increase contrast to improve readability.

- **Captions and transcriptions**: For multimedia content, captions and transcriptions make audio and video accessible to hearing-impaired users.

Benefits of Inclusive Design

Inclusive design benefits all users, not just those with disabilities. For example, ensuring that buttons are large enough to be clicked easily can help users with motor impairments and make the interface more accessible to users on mobile devices. By designing for the full spectrum of users, software becomes more usable and accessible to everyone.

Usability in Different Software Types

Usability considerations vary depending on the type of software being developed. While the principles of usability remain the same, different software types may prioritize certain aspects of usability over others.

Web Applications

Web applications often focus on learnability, responsiveness, and error prevention, as users expect to perform tasks quickly and without complications. Browser compatibility, page load speed, and mobile responsiveness are critical aspects of usability for web applications.

Mobile Applications

Mobile apps must account for touch interfaces, smaller screen sizes, and limited input options. Usability in mobile apps prioritizes clear navigation, large touch targets, and efficient workflows that minimize the number of steps required to complete tasks.

Enterprise Software

Enterprise software systems are often complex, with a wide range of features and functions. Usability in this context focuses on efficiency, scalability, and error tolerance, as the software is typically used by professionals who need to perform tasks quickly and accurately. Usability improvements in enterprise software can directly impact productivity and operational efficiency.

Consumer Software

For consumer-facing software, user satisfaction and emotional appeal are key aspects of usability. The design must be intuitive, visually appealing, and enjoyable to use, ensuring that users remain engaged and continue using the software.

Common Usability Challenges

Designing highly usable software is not without its challenges. Common obstacles include balancing simplicity with functionality, managing user diversity, and addressing technical limitations.

Balancing Simplicity and Functionality

Striking the right balance between simplicity and feature-richness is a major challenge in usability design. While users generally prefer simple interfaces, overly simplified designs may lack the advanced features that experienced users need. Designers must find ways to offer flexibility, such as hiding advanced features in menus or offering customizable settings to accommodate both novice and expert users.

Managing User Diversity

Users vary in their technical expertise, preferences, and needs. Designing for a broad audience can be challenging, as what works well for one group of users may be confusing or inefficient for another. Creating customizable interfaces, user profiles, or adaptable workflows can help address this challenge.

Addressing Technical Constraints

Technical limitations, such as slow network connections, device restrictions, or browser compatibility issues, can hinder usability. Designing software to work across various devices, operating systems, and environments requires careful planning and testing.

Usability Metrics and Measurement

Measuring usability involves collecting data to assess how well users can interact with the software and how satisfied they are with their experience. Usability metrics provide quantitative insights into the effectiveness of the design and help identify areas for improvement.

Common Usability Metrics

- **Task success rate**: Measures the percentage of tasks that users can complete successfully.

- **Time on task**: Tracks how long it takes users to complete a task, indicating efficiency.

- **Error rate**: Counts the number of errors users make while interacting with the system.

- **Satisfaction scores**: Gathers user feedback on their satisfaction with the software, often through surveys or rating scales.

- **Learnability**: Measures how quickly new users can become proficient with the software.

Using Metrics to Improve Usability

Usability metrics can be used to track improvements over time and ensure that changes to the software enhance the user experience. For example, if task success rates increase after a redesign, it suggests that the changes have improved usability. Continuous monitoring of these metrics helps maintain high usability standards throughout the software's lifecycle.

Conclusion

In today's competitive software landscape, usability is a critical factor that can make or break a product's success. It directly affects user satisfaction, adoption rates, productivity, and the overall user experience. By understanding the key usability concepts—such as learnability, efficiency, and error tolerance—designers and developers can create software that effectively meets user needs.

This chapter has introduced the essential components of usability, explored the benefits of high usability, and outlined best practices for designing user-friendly software. By prioritizing usability from the early stages of development and continuously refining the user interface through testing and feedback, software teams can deliver products that are not only functional but also intuitive, efficient, and enjoyable to use. In an increasingly user-centric world, investing in usability is essential for building software that stands out and drives success.

CHAPTER 2

Principles of Effective Software Design

Software design is a key element of the software development lifecycle (SDLC), shaping how requirements are translated into a technical solution that is efficient, scalable, and maintainable. Effective software design bridges the gap between user needs and system implementation by applying a structured approach to managing complexity. The principles that govern effective software design emphasize clarity, modularity, reusability, and flexibility.

Design decisions taken early in the development process often have long-term consequences, affecting the system's ease of maintenance, scalability, performance, and security. This chapter delves into the principles that help developers create software that stands the test of time, adapting to changing requirements and evolving technologies.

Software design is a blend of art and science. It requires balancing technical constraints with the demands of users and stakeholders. The focus here is on object-oriented design principles, but many of these concepts are applicable across paradigms, from procedural to functional programming. Ultimately, well-designed software is not only about solving current problems but anticipating future challenges in an efficient and sustainable way.

© Saurav Bhattacharya 2025
P. Gujar, *Data Usability in the Enterprise*, https://doi.org/10.1007/979-8-8688-1183-8_2

Design Fundamentals

Clarity and Simplicity

One of the most fundamental principles of software design is **clarity**, which refers to how easily code can be understood by humans, not just machines. Readable code reduces errors and facilitates future enhancements. Writing clear code involves using descriptive variable names, consistent formatting, and avoiding complex or obscure language features unless absolutely necessary.

Simplicity complements clarity. Complex systems are harder to maintain and more prone to errors. The goal should be to build software systems that are as simple as possible while meeting all functional requirements. Simplicity is not just about reducing the lines of code but also about reducing unnecessary dependencies and designing straightforward and easy-to-reason algorithms.

Reusability

Designing software components that can be reused across different contexts is crucial for efficiency. Reusable code reduces redundancy, accelerates development in future projects, and enhances consistency across systems. Do the following to maximize reusability.

- Avoid tightly coupling code to specific implementations.

- Rely on interfaces or abstract classes to provide flexibility.

- Write generic functions or classes that work with a variety of inputs or types.

By keeping reusability in mind, developers create systems that scale and evolve with minimal rework.

Iterative Design

Effective software design is rarely a linear process. Instead, it is iterative. Designers should expect to revisit and refine their designs as they receive feedback, encounter performance challenges, or adapt to changing requirements. Agile methodologies emphasize the importance of iterative design by allowing continuous feedback from stakeholders, helping teams adjust as they progress.

Iteration also plays a role in handling complexity. Rather than attempting to design the entire system upfront, start with a simple design and evolve it over time as new requirements emerge.

Modularity

Modularity is the process of dividing a system into discrete components or modules that can be developed, tested, and maintained independently. This principle is essential in both small-scale applications and large-scale enterprise systems.

Benefits of Modularity

- **Maintainability:** When a system is broken into modules, individual modules can be maintained, updated, or replaced without impacting other parts of the system.

- **Reusability**: Modules can be reused across multiple projects, minimizing redundancy and speeding up development.

- **Parallel development**: Different teams or developers can work on separate modules concurrently, improving development speed and team efficiency.

- **Testing**: Testing is more efficient when focusing on small, modular units of functionality. Unit tests can be created for individual modules to validate their behavior in isolation.

Best Practices for Modularity

Modules should have a single, clearly defined purpose, with a well-defined interface for interacting with other modules. Communication between modules should be limited to these interfaces to avoid unintended dependencies or side effects. Keeping modules loosely coupled and highly cohesive will lead to a more flexible and adaptable system.

Abstraction

Abstraction allows developers to focus on the higher-level functionality of a system while hiding the underlying complexity. It plays a pivotal role in managing complexity in large systems by breaking them down into more manageable layers or components.

Data Abstraction

Data abstraction is achieved through encapsulation (discussed further in the next section) by hiding the internal representation of data from the user and exposing only the necessary operations. For example, in an object-oriented language, a class can encapsulate data and provide getter and setter methods that hide the internal data structure from external classes. This reduces the likelihood of errors caused by unintended interactions with data.

Control Abstraction

Control abstraction simplifies the control flow of a program by allowing developers to use high-level constructs like functions, loops, and conditional statements without needing to understand the underlying mechanics. This principle allows code to be structured more clearly, reducing complexity and making it easier to understand.

By abstracting both data and control mechanisms, developers can reduce the cognitive load when working with complex systems. Each layer of abstraction should focus on solving a specific type of problem while hiding unnecessary details from other parts of the system.

Encapsulation

Encapsulation is the practice of bundling data and the methods that operate on that data into a single entity, such as a class, and restricting access to some of the object's components. This principle not only protects the integrity of the data but also defines clear boundaries between different components in a system.

The following are some of the benefits of encapsulation.

- **Data protection**: By controlling access to data, encapsulation prevents external classes from modifying it in ways that could break the system.

- **Reduced complexity**: Encapsulation hides internal complexities, making the system easier to use and understand from the outside.

- **Maintainability**: When the internal implementation of a class changes, encapsulation ensures that external components do not need to be modified, provided the public interface remains unchanged.

Encapsulation is foundational to object-oriented programming (OOP) and directly contributes to modularity, as it allows for a clear separation of concerns between different parts of the system.

Separation of Concerns

Separation of concerns is a design principle that divides a software system into distinct sections, each responsible for a separate concern or aspect of the system's functionality. This separation improves modularity, making it easier to develop, maintain, and evolve the system.

In a web application, the system can be divided into three layers.

- **Presentation layer**: Manages user interaction and interface display.

- **Business logic layer**: Handles application-specific rules and logic.

- **Data access layer**: Manages data storage and retrieval.

By clearly separating these concerns, developers can make changes to the presentation without affecting the business logic or data access and vice versa.

Coupling and Cohesion

Coupling and cohesion are critical concepts in software design that evaluate how well software components are connected and how focused they are on specific tasks.

Coupling

Coupling refers to the degree of dependency between different modules. Loose coupling means that modules are independent of each other, reducing the impact of changes in one module on another. Tight coupling indicates that modules are highly dependent, making the system more fragile and harder to modify.

Loose coupling is desirable in most cases because it allows for more flexible and maintainable systems. The following are some strategies for achieving loose coupling.

- Using interfaces and abstract classes

- Employing design patterns

- Minimizing shared resources or global states

Cohesion

Cohesion measures how closely related the responsibilities of a module or component are. **High cohesion** means that a module performs a single task or closely related tasks, making the code easier to understand and maintain. **Low cohesion** means that a module is trying to do too many unrelated things, leading to a more complex and error-prone system.

A well-designed system has high cohesion and low coupling, promoting better modularity and maintainability.

Design Patterns

Design patterns are standardized solutions to common design problems in software development. They provide a blueprint for solving specific types of issues that developers often encounter. Design patterns are not code but templates that can be applied in a variety of programming languages and contexts.

Design Pattern Categories

Design patterns are generally classified into three categories.

- **Creational patterns** deal with object creation mechanisms, optimizing flexibility, and reusing existing code. The following are two examples.

  - **Singleton**: Ensures a class has only one instance and provides a global point of access to it.

  - **Factory method**: Defines an interface for creating objects but lets subclasses alter the type of objects that will be created.

- **Structural patterns** deal with object composition, simplifying relationships between different entities. The following are two examples.

  - **Adapter**: Allows incompatible interfaces to work together by providing a "wrapper" class.

  - **Composite**: Organizes objects into tree structures to represent part-whole hierarchies.

- **Behavioral patterns** focus on object interaction and responsibility. The following are two examples.

  - **Observer**: Defines a dependency between objects so that when one object changes state, its dependents are notified.

  - **Strategy**: Defines a family of algorithms and allows their interchangeability without altering client code.

Benefits of Using Design Patterns

- **Reusability**: Patterns are proven solutions that can be reused across different projects, speeding up development.

- **Maintainability**: Patterns provide a clear structure for solving design problems, making code easier to understand and modify.

- **Scalability**: Many patterns support scalability by decoupling system components or optimizing how resources are managed.

While design patterns offer many benefits, it is essential to use them judiciously. Overuse of patterns can lead to unnecessary complexity.

SOLID Principles

The SOLID principles, coined by Robert C. Martin, provide a framework for designing flexible, maintainable, and scalable software. These principles focus on creating systems that can adapt to change and are easier to manage over time.

- **Single responsibility principle (SRP)**: A class should have only one reason to change, meaning it should only perform one task or represent one concept. This principle helps reduce the impact of changes, as modifications to one functionality should not affect unrelated functionalities.

- **Open/closed principle (OCP)**: Software entities should be open for extension but closed for modification. This principle encourages developers to build systems that can be extended with new features without altering existing code. For example, by using interfaces or abstract classes, new functionality can be added without changing the code that relies on the original interface.

- **Liskov Substitution Principle (LSP)**: Objects of a superclass should be replaceable with objects of a subclass without affecting the correctness of the program. This principle ensures that subclass objects behave consistently with the expectations set by their parent class, promoting flexibility in code reuse.

- **Interface segregation principle (ISP)**: Clients should not be forced to depend on interfaces they do not use. Instead of creating large, monolithic interfaces, break them into smaller, more specific ones. This allows clients to depend only on the functionalities they need, reducing unnecessary dependencies.

- **Dependency inversion principle (DIP)**: High-level modules should not depend on low-level modules; both should depend on abstractions. This principle is often achieved through dependency injection, where objects are provided their dependencies from an external source rather than creating them internally. By doing so, systems are more decoupled and easier to modify.

The SOLID principles promote better design practices, leading to more maintainable, scalable, and flexible software architectures.

Performance and Scalability

Designing for performance and scalability is essential in modern software development, especially for applications that handle increasing user or data loads. Effective software design addresses both the immediate needs for fast and responsive applications and the long-term requirements for growth.

Performance Optimization

Performance refers to how efficiently a system executes tasks. While developers should avoid premature optimization, it is crucial to design systems that can perform well under realistic conditions. The following are some considerations.

- **Algorithmic efficiency**: Choose appropriate data structures and algorithms that minimize time complexity.

- **Caching**: Store frequently accessed data in a cache to reduce the time required to retrieve it from slower storage systems.

- **Concurrency**: Enable systems to handle multiple tasks simultaneously using threading, asynchronous processing, or parallel execution.

- **Minimizing resource use**: Optimize memory and CPU usage, avoiding memory leaks and unnecessary computation.

Scalability Strategies

Scalability refers to a system's ability to handle increasing amounts of work, whether by adding more resources or optimizing the existing ones. The following are common strategies.

- **Horizontal scaling**: Adding more machines to handle an increased load (e.g., adding more servers to a web application cluster).

- **Vertical scaling**: Increasing the resources (CPU, memory, storage) of an existing machine to improve performance.

- **Load balancing**: Distributing incoming requests across multiple servers to prevent any single machine from becoming overwhelmed.

- **Database sharding**: Splitting a database into smaller, more manageable pieces to distribute the load more evenly.

Designing for scalability requires considering the system's future growth potential and ensuring that adding more users, data, or resources will not degrade performance.

Usability and User-Centered Design

While much of software design focuses on the underlying architecture, usability is equally important. Usability measures how easy and efficient it is for users to accomplish their goals using the system. User-centered design ensures that the design process focuses on users' needs and behaviors.

Key Usability Principles

- **Consistency**: Interfaces should have consistent behavior and appearance throughout the application. Consistency helps users form mental models and reduces the learning curve.

- **Feedback**: Providing users with immediate feedback about their actions helps them understand the system's state. For example, showing a progress bar during file uploads informs users that their action is being processed.

- **Affordances**: Design elements should suggest their functionality. Buttons should look clickable, and input fields should indicate where users can type.

- **Simplicity**: Avoid unnecessary complexity in user interfaces. Keep controls intuitive and task-specific to minimize the cognitive load on the user.

- **Accessibility**: Ensure that people with disabilities can use the software. This includes support for screen readers, keyboard shortcuts, and adjustable text sizes.

User-Centered Design Process

UCD involves understanding the end users' needs early in the design process and incorporating their feedback throughout development. The key steps include the following.

- **User research**: Interviewing and observing users to understand their goals, pain points, and behaviors.

- **Prototyping**: Developing wireframes or low-fidelity prototypes to visualize design ideas and get early feedback from users.

- **Usability testing**: Testing the software with real users to identify usability issues and improve the design.

- **Iterative design**: Continuously refining the design based on user feedback and testing results.

By focusing on usability and user-centered design, developers can create software that is not only functional but also pleasant and efficient to use.

Maintainability and Extensibility

Maintainability and extensibility are essential qualities of long-lived software systems. **Maintainability** refers to how easily a system can be understood, corrected, and enhanced. **Extensibility** refers to how easily new features can be added without disrupting the existing system.

Maintainability

Developers should do the following to improve maintainability.

- **Write clean code.** Adhere to coding conventions and best practices to ensure the code is easy to read and understand.

- **Document code and system architecture.** Well-documented code helps developers quickly grasp the purpose and structure of the codebase. Similarly, documenting the system architecture helps new developers understand how different components interact.

- **Refactor regularly.** Refactoring improves the internal structure of the code without changing its external behavior. It helps eliminate technical debt and keeps the codebase clean and efficient.

- **Use modular design.** Modularity makes it easier to isolate and fix issues, improving maintainability.

Extensibility

Developers should do the following to improve extensibility.

- **Use abstractions.** Designing with interfaces, abstract classes, and other forms of abstraction helps keep the system flexible and open to future changes.

- **Design for change.** Anticipating future requirements during the design phase can reduce the need for major overhauls later. For example, building systems with extensible plug-ins or APIs allows for easier feature expansion.

- **Apply SOLID principles.** The SOLID principles, especially the open/closed principle, support extensibility by allowing developers to add new features without altering existing code.

By focusing on maintainability and extensibility, developers ensure that their software can evolve over time while remaining functional and efficient.

Testing and Debugging

Testing and debugging are integral parts of the software development process. They ensure that the software meets the desired requirements and functions correctly under various conditions. A well-designed system supports effective testing and debugging by being modular, maintainable, and easy to understand.

Types of Testing

- **Unit testing** verifies that individual components or units of the system work as expected. Unit tests are typically automated and focus on small, isolated pieces of code.

- **Integration testing** ensures that different modules or components of the system work together correctly. Integration testing verifies that interfaces between modules are functioning as expected.

- **System testing** evaluates the system as a whole, checking that all components work together to meet the overall requirements.

- **Acceptance testing** confirms that the system meets the business requirements and is ready for release. This type of testing is often conducted by end-users or stakeholders.

- **Regression testing** ensures that changes to the code, such as bug fixes or new features, do not introduce new issues. Automated regression testing is especially useful for large systems.

Test-Driven Development

Test-driven development (TDD) is a software development methodology in which tests are written before the code. The following describes the steps of a typical TDD cycle.

1. Write a test for a new feature or functionality.

2. Run the test and watch it fail (since the feature has not yet been implemented).

3. Write the minimal amount of code needed to make the test pass.

4. Refactor the code to improve its structure without altering functionality.

5. Repeat the process for additional features.

TDD ensures that every new feature is backed by tests, leading to higher-quality, more reliable software.

Debugging

Effective debugging requires tools and techniques that help identify and fix issues quickly.

- **Logging and monitoring**: Implement logging throughout the application to capture detailed information about its runtime behavior. Monitoring tools can help track system performance and detect anomalies.

- **Breakpoints**: Debugging tools allow developers to set breakpoints, which pause the execution of a program at a specific point. This helps developers inspect the state of the system and identify where errors occur.

- **Automated error reporting**: When errors occur in production, automated error reporting systems can capture details about the environment, stack traces, and user actions leading up to the error. This information helps developers diagnose and fix issues more efficiently.

Testing and debugging are iterative processes that, when done correctly, lead to more reliable, stable software.

Security in Design

Security is an increasingly critical aspect of software design, especially in an era where data breaches and cyberattacks are commonplace. Incorporating security into the design process ensures that software systems are robust against malicious attacks and unintended vulnerabilities.

Common Security Principles

- **Least privilege**: Systems should grant users and components only the minimum level of access necessary to perform their tasks. This reduces the attack surface and minimizes the impact of security breaches.

- **Defense in depth**: Instead of relying on a single security mechanism, implement multiple layers of defense. For example, in a web application, you might combine encryption, authentication, access control, and intrusion detection to protect sensitive data.

- **Input validation**: Always validate input from users or external systems to prevent attacks like SQL injection, cross-site scripting (XSS), or buffer overflows. Input validation ensures that only properly formatted data is processed by the system.

- **Encryption**: Sensitive data, whether at rest or in transit, should be encrypted. Encryption ensures that even if data is intercepted or stolen, it cannot be easily read or used.

- **Auditing and logging**: Maintain detailed logs of system activities, including user actions, authentication attempts, and system errors. Logs are crucial for detecting, investigating, and responding to security incidents.

- **Regular security updates**: Keep all software components current, including third-party libraries and frameworks. Many security vulnerabilities are caused by outdated or unpatched software.

Secure Software Development Lifecycle

A secure software development lifecycle (SSDLC) integrates security practices into each phase of development, from design to deployment. The following are some key steps.

- **Threat modeling**: Identify potential threats and vulnerabilities early in the design phase. Threat modeling helps developers anticipate security risks and design systems to mitigate them.

- **Security testing**: Conduct security testing throughout development, including static code analysis, penetration testing, and vulnerability scanning.

- **Post-release monitoring**: Continuously monitor the system after release for new vulnerabilities, performance issues, and attacks. Implement security patches promptly to address newly discovered vulnerabilities.

By incorporating security considerations into the design process, developers can create systems that are more resilient against attacks and better protect user data.

Documentation and Communication

Good software design is not just about writing code—it also involves clear communication and documentation. Documentation serves as a guide for future developers, stakeholders, and users, ensuring that the system can be understood, maintained, and extended effectively.

Types of Documentation

- **Technical documentation**: This includes documentation of the system's architecture, design decisions, data models, APIs, and configuration details. It helps developers understand how the system is structured and how its components interact.

- **User documentation**: Manuals or guides for end-users that explain how to use the software. Clear user documentation improves the user experience and reduces the need for support.

- **Code comments**: Inline comments within the codebase explain complex logic, describe non-obvious decisions, or outline areas that may need attention in the future.

- **Test documentation**: Tests should be well-documented so that other developers can understand the purpose and expected behavior of each test case.

Importance of Communication

Effective communication within a development team ensures everyone is aligned on the system's goals and architecture. This includes the following.

- **Design meetings**: Regular design meetings allow teams to discuss design choices, share feedback, and collaboratively solve challenges.

- **Code reviews**: Peer reviews of code help catch bugs, improve code quality, and ensure adherence to design principles and standards.

- **Cross-team collaboration**: In larger organizations, teams working on different modules or layers must collaborate to ensure that their components integrate smoothly and meet the overall system objectives.

By prioritizing clear documentation and open communication, software teams can build more coherent, maintainable, and successful systems.

Refactoring

Refactoring is the process of restructuring existing code without changing its external behavior. It aims to improve the code's structure, readability, and maintainability. Refactoring is crucial to long-term software development because it helps eliminate technical debt and keeps the codebase clean and efficient.

The following are key reasons for refactoring.

- **Improves code readability**: Over time, code can become cluttered or difficult to understand. Refactoring simplifies and clarifies the codebase.

- **Removes redundancy**: Duplicate code increases the risk of errors and inconsistencies. Refactoring consolidates redundant code into reusable components.

- **Enhances performance**: Refactoring can optimize algorithms or data structures to improve system performance without altering functionality.

- **Adapting to new requirements**: As business needs change, refactoring helps evolve the system to meet new requirements while maintaining code quality.

Common Refactoring Techniques

The following describes some common refactoring techniques.

- **Extract a method.** Break down long or complex methods into smaller, more manageable pieces, each with a clear, distinct purpose.

- **Rename variables or functions.** Use descriptive names for variables, methods, or classes to make the code easier to understand.

- **Simplify conditional logic.** Refactor complex if-else or switch-case statements into simpler, more readable structures.

- **Encapsulate fields.** Ensure that fields within a class are accessed through getter and setter methods rather than directly, enhancing flexibility.

- **Inline temporary variables.** If a temporary variable is unnecessary, inline it to simplify the code.

Regular refactoring improves code quality over time, reduces the risk of bugs, and ensures the system remains adaptable to future changes.

Version Control and Continuous Integration

Version control and continuous integration are essential practices for modern software development. They help teams collaborate effectively, manage changes to the codebase, and ensure the system remains stable and reliable during development.

Version Control

Version control systems (VCS), such as Git, track changes to the codebase, which allows developers to do the following.

- **Collaborate**: Multiple developers can work on the same codebase simultaneously, with the ability to merge their changes.

- **Revert changes**: If a change introduces a bug or breaks the system, developers can easily revert to a previous version of the code.

- **Track history**: VCS keeps a history of all changes, providing a detailed audit trail of who made changes, when, and why.

The following are effective version control strategies.

- **Branching**: Use branches to isolate new features or bug fixes from the main codebase. Once the work is complete, the branch can be merged back into the main codebase after review.

- **Tagging**: Tagging specific versions of the codebase (e.g., for releases) helps track stable versions and identify the state of the system at a given point in time.

Continuous Integration

Continuous integration (CI) is the practice of automatically building and testing code whenever changes are made, ensuring that the system remains stable and functional throughout development. The following are some key CI practices.

- **Automated builds**: Every time a change is committed to the version control system, the CI server automatically builds the system to ensure that it compiles successfully.

- **Automated testing**: CI servers run unit, integration, and system tests to ensure that new changes do not break existing functionality.

- **Code quality checks**: CI tools can automatically check code for adherence to coding standards, detect potential issues, and enforce style guides.

By integrating CI with version control, teams can catch issues early, reduce the risk of introducing bugs, and maintain a high level of code quality throughout the development process.

Conclusion

The principles of effective software design—modularity, abstraction, encapsulation, separation of concerns, and adherence to best practices such as SOLID—are vital for building systems that are robust, maintainable, and scalable. By following these principles, developers can create software that not only meets current requirements but is also flexible enough to evolve over time.

Design is an iterative process that requires constant feedback, testing, and refinement. As technologies and business needs change, the ability to adapt and improve a system without causing disruptions becomes crucial. Principles such as performance optimization, security, testing, and refactoring ensure that systems remain efficient, secure, and maintainable.

In summary, effective software design is about managing complexity, facilitating collaboration, and ensuring that systems are built to last. By applying the principles outlined in this chapter, developers can create high-quality software that delivers value to users and stands the test of time.

CHAPTER 3

Enhancing Software Accessibility for Use Cases

Software accessibility is an essential component of modern software design. It aims to ensure that all users, regardless of their physical, sensory, or cognitive abilities, can fully interact with and benefit from digital tools and platforms. As the world becomes increasingly reliant on digital solutions for everyday tasks, from education and healthcare to social networking and e-commerce, accessibility becomes a critical factor in shaping the inclusivity of our digital landscape.

The demand for accessible software is driven not only by the growing awareness of disability rights but also by legal requirements and global standards that mandate accessibility for digital products and services. However, accessibility is not just about compliance; it's about creating equitable user experiences for everyone. This chapter explores the key principles of accessible software design, practical strategies for implementation, and real-world examples, providing a comprehensive guide for developing software that meets the needs of diverse users.

© Saurav Bhattacharya 2025
P. Gujar, *Data Usability in the Enterprise*, https://doi.org/10.1007/979-8-8688-1183-8_3

The chapter also examines how the intersection of accessibility and usability creates inclusive experiences that benefit all users, not just those with disabilities. From universal design principles to the role of assistive technologies, this chapter covers everything you need to know about enhancing software accessibility across a variety of use cases and contexts.

Understanding Accessibility in Software Design

Accessibility in software design refers to creating digital products and services that are usable by everyone, including people with disabilities. Accessibility covers a wide range of conditions, from visual and hearing impairments to mobility issues, cognitive disabilities, and even situational limitations, such as using a device in low-light environments or with limited connectivity.

Defining Disability and Accessibility

According to the World Health Organization (WHO), disabilities are diverse and multidimensional, affecting people in different ways depending on their environment. For example, a person with a hearing impairment might face barriers in environments that rely heavily on audio cues, and someone with a visual impairment could struggle with non-accessible websites or apps.

When applied to software, accessibility addresses these barriers by ensuring the following.

- Information is **perceivable** by all users, regardless of sensory abilities.

- User interfaces are **operable**, providing alternatives for navigation and interaction.

- Content is **understandable**, making it simple for users with cognitive disabilities to comprehend.

- Digital experiences are **robust** and compatible with a wide variety of devices and assistive technologies.

The Importance of Accessibility for Different Use Cases

Accessibility is relevant to all software, but certain use cases highlight the critical importance of designing with diverse abilities in mind. From online education platforms to e-commerce and healthcare systems, accessible software ensures equal access to vital services and opportunities for all users.

Education

Education platforms are a primary example where accessibility is essential. Digital classrooms, learning management systems, and other educational tools need to be designed to accommodate a range of learning needs, including those with disabilities.

- **Students with visual impairments** benefit from screen reader-friendly interfaces and high-contrast options.

- **Deaf or hard-of-hearing students** rely on captions for video lectures and transcripts for audio materials.

- **Students with cognitive impairments** require simple navigation, consistent layouts, and additional scaffolding to assist comprehension.

Inaccessible education software can exclude students from crucial learning opportunities, undermining the goal of inclusive education. On the other hand, accessible tools support diverse learning styles, enhance comprehension, and create more engaging educational experiences.

Healthcare

Accessibility in healthcare software ensures that patients and providers alike can interact with medical records, appointment systems, and telehealth platforms regardless of physical or cognitive limitations. Accessible healthcare tools allow the following.

- **Patients with mobility impairments** can schedule appointments or consult with doctors via accessible online portals.

- **Providers with disabilities** can access electronic health records (EHRs) and participate in digital workflows without unnecessary barriers.

Ensuring that healthcare software is accessible not only enhances the quality of care for all patients but also fosters inclusivity within healthcare workforces.

E-commerce

In e-commerce, accessibility plays a crucial role in enabling all customers to navigate online stores, compare products, and complete transactions. Accessible e-commerce websites and apps do the following.

- Support **keyboard navigation** for users who cannot use a mouse

- Provide **alternative text for images** so that screen reader users can understand product images

- Offer **accessible forms** for checkout processes, making it easy for all users to complete purchases

Excluding users with disabilities from online commerce not only limits their ability to shop independently but also reduces a company's market reach. Accessibility in e-commerce enhances user satisfaction and broadens customer bases.

Universal Design Principles

Universal design is the concept of designing products, environments, and systems that all people can use to the greatest extent possible without the need for adaptation or specialized design. In software development, universal design ensures that accessibility is embedded into the core of the product, benefiting all users, not just those with disabilities.

The Seven Principles of Universal Design

- **Equitable use**: The design should be usable by people with diverse abilities. This means creating interfaces that do not disadvantage or stigmatize any group of users.

- **Flexibility in use**: The software should accommodate a wide range of individual preferences and abilities, allowing users to choose how they interact with the system. For example, providing multiple input methods—such as touch, voice, or keyboard—offers flexibility for users with different physical abilities.

- **Simple and intuitive use**: Regardless of the user's experience or ability, the design should be easy to understand. This includes eliminating unnecessary complexity and providing clear instructions.

51

- **Perceptible information**: The design should communicate necessary information to the user, regardless of ambient conditions or the user's sensory abilities. This might involve offering text alternatives for audio cues or making text legible in low-contrast environments.

- **Tolerance for error**: The design should minimize the potential for accidents and unintended actions, as well as provide easy ways to recover from errors.

- **Low physical effort**: The design should allow for efficient, comfortable use, minimizing repetitive actions or sustained physical effort.

- **Size and space for approach and use**: The design should ensure adequate space for users to interact with the interface, taking into account mobility aids or physical limitations.

Universal design principles serve as a foundation for creating products that are accessible to the widest possible audience while reducing the need for individual accommodations.

Key Accessibility Standards and Guidelines

To create accessible software, developers must follow established accessibility standards and guidelines that ensure compliance with legal requirements and best practices. These frameworks provide clear, actionable criteria for making digital products accessible to all users, regardless of their abilities.

Web Content Accessibility Guidelines

The Web Content Accessibility Guidelines (WCAG), developed by the World Wide Web Consortium (W3C), are the most widely recognized standards for digital accessibility. WCAG is structured around four core principles.

- **Perceivable**: Information must be presented in a way that users can perceive, whether visually, audibly, or through alternative formats.

- **Operable**: Users must be able to interact with the interface and navigate content, whether by keyboard, mouse, or assistive technology.

- **Understandable**: Content and interfaces should be easy to comprehend, regardless of cognitive ability.

- **Robust**: Digital content must be compatible with current and future assistive technologies.

WCAG is divided into three conformance levels.

- **Level A** (minimum): Basic web accessibility features.

- **Level AA** (mid-range): Deals with the biggest and most common barriers for users with disabilities.

- **Level AAA** (maximum): The highest level of web accessibility.

Most legal requirements globally reference WCAG 2.0 or WCAG 2.1 at Level AA conformance, though some organizations aim for Level AAA for maximum inclusivity.

Section 508

In the United States, Section 508 of the Rehabilitation Act mandates that federal agencies and organizations receiving federal funds ensure their electronic and information technology is accessible to people with disabilities. Section 508 closely aligns with WCAG standards but includes specific provisions for software, documents, and multimedia used by government agencies.

ISO 9241-171

The ISO 9241-171 standard focuses on the ergonomics of human-system interaction, providing specific guidance on software accessibility. This international standard covers the design of user interfaces, ensuring they are accessible to people with physical, sensory, and cognitive disabilities.

EN 301 549

In the European Union, EN 301 549 is the standard for ensuring the accessibility of information and communication technologies (ICT), including websites, mobile apps, and digital services. It references WCAG 2.1 and outlines requirements for public sector bodies to ensure their digital tools are accessible to all citizens.

By adhering to these standards, developers can create software that is not only accessible to users with disabilities but also legally compliant and ready for global distribution.

Inclusive Design: Bridging Usability and Accessibility

While accessibility focuses on removing barriers for users with disabilities, inclusive design is a broader methodology that considers the full range of human diversity, including age, culture, gender, and abilities. Inclusive design ensures that products work well for as many people as possible and that users can personalize their experiences based on their preferences and needs.

Co-Designing with Users

Inclusive design starts with understanding the users and their needs. By involving users with diverse abilities and backgrounds in the design process, developers can create software that reflects real-world experiences and challenges. This practice, known as **co-design**, allows for more accurate and practical solutions to accessibility issues.

Adaptive and Flexible Solutions

In inclusive design, flexibility is key. Developers should offer multiple ways to complete tasks or interact with the software, allowing users to choose the method that works best for them. For example, a user interface might allow users to do the following.

- Navigate using voice commands, touch, or a keyboard

- Adjust the size, contrast, or color scheme of the text to suit individual preferences

- Personalize the layout to prioritize the most frequently used features

This adaptability ensures that the software is inclusive and meets the needs of a diverse user base.

Personalization

Inclusive design encourages personalization by allowing users to customize their interface based on their preferences and abilities. For example, users with visual impairments might prefer a high-contrast mode, while users with cognitive disabilities might benefit from simplified interfaces with fewer distractions.

By building personalization options into the software, developers can accommodate a wide range of user needs without requiring extensive individual adaptations.

Designing for Users with Visual Impairments

Designing software for users with visual impairments involves ensuring that all visual information can be accessed in alternative ways, such as through audio, tactile feedback, or screen reader technology. Visual impairments range from color blindness and low vision to total blindness, and each presents unique challenges in software design.

Screen Reader Compatibility

Screen readers like NVDA (NonVisual Desktop Access) and JAWS (Job Access With Speech) convert text and other on-screen elements into speech or braille output, allowing users with visual impairments to interact with digital content. To ensure that software is compatible with screen readers, developers should do the following.

- Use semantic HTML to provide structure and meaning to the content.

- Add **alt text** to all images, ensuring that screen readers can describe visual content to users.

- Use ARIA (Accessible Rich Internet Applications) labels and roles to define the purpose of interactive elements, such as buttons and forms.

By ensuring that all on-screen elements are properly labeled and described, developers can create software that is fully navigable using screen readers.

High-Contrast and Color Adjustments

For users with low vision or color blindness, color contrast can be a major accessibility barrier. To address this, developers should do the following.

- Use high-contrast color schemes that make text and interactive elements easier to distinguish.

- Provide customizable contrast and color options, allowing users to adjust the interface based on their individual preferences.

WCAG guidelines recommend a contrast ratio of at least 4.5:1 for normal text and 3:1 for large text.

Text Resizing and Zooming

Users with low vision often rely on text resizing or screen magnification to read on-screen content. Developers should ensure that text can be resized without breaking the layout of the interface. Additionally, elements should remain functional when zoomed in, and users should be able to navigate the software even at higher magnification levels.

Keyboard Navigation

For many users with visual impairments, navigating a software interface using only a keyboard is essential. Developers should ensure that all interactive elements are accessible via keyboard navigation, with clear focus indicators showing which element is currently active. Proper tab ordering is crucial, as it helps users navigate through the interface in a logical sequence.

Designing for Users with Hearing Impairments

Users with hearing impairments rely on visual cues and text-based alternatives to access auditory information. Designing software that accommodates these needs involves providing accessible alternatives for audio content and ensuring that auditory alerts are supplemented with visual indicators.

Subtitles and Captions for Multimedia

Video and audio content should include subtitles or closed captions to ensure that users with hearing impairments can access the information being presented. Subtitles should not only transcribe dialogue but also include descriptions of non-verbal sounds, such as music, laughter, or sound effects.

For audio-only content, such as podcasts or voice messages, developers should provide **transcripts** that users can read as an alternative to listening.

Visual Alerts

Auditory alerts, such as system sounds or notification chimes, should be accompanied by visual alerts to ensure that users with hearing impairments are aware of important events. Visual alerts can take the form of pop-up notifications, flashing icons, or color changes, ensuring that the information is conveyed through visual means.

Real-Time Text Communication

In applications that rely on real-time communication, such as video conferencing or customer support chat, providing real-time text (RTT) options can enhance accessibility for users who are deaf or hard of hearing. RTT allows users to type and read messages in real time, creating an alternative to voice-based communication.

Assistive Technologies for Hearing Impairments

Some users with hearing impairments rely on assistive technologies, such as hearing aids or cochlear implants, that interface with digital devices. Developers should ensure that their software supports these technologies by providing options for audio streaming, speech-to-text transcription, or direct connectivity with hearing aids.

Designing for Users with Motor Impairments

Motor impairments can make it difficult for users to interact with software using traditional input methods like a mouse or keyboard. Designing for users with motor impairments involves providing alternative input options and simplifying interactions to reduce physical strain.

Keyboard Accessibility

For users with limited mobility, keyboard navigation is often the primary method of interacting with software. Developers should ensure that all interactive elements are accessible using keyboard shortcuts, tab navigation, and key commands. Focus indicators should be visible and clearly show which element is currently selected.

Voice Control

Voice control can provide an alternative input method for users with motor impairments who cannot use a mouse or keyboard. By integrating voice recognition technologies, such as Google's Speech-to-Text API or Apple's SiriKit, developers can allow users to control the software using voice commands.

Customizable Input Sensitivity

For users with limited fine motor skills, adjusting the sensitivity of input devices, such as touchscreens or mice, can make interactions easier. Providing options to customize the speed, accuracy, and pressure required for touch or click actions can reduce physical effort and enhance accessibility.

Alternative Input Devices

Some users with severe motor impairments rely on alternative input devices, such as adaptive keyboards, switch systems, or eye-tracking devices, to control their computers. Ensuring compatibility with these devices is essential for accessibility. Developers should test their software using these alternative input methods and ensure that the interface remains fully functional.

Cognitive Disabilities and Software Design

Cognitive disabilities encompass a wide range of conditions, from learning disabilities and attention disorders to memory impairments and autism spectrum disorders. Designing software that is accessible to users with cognitive disabilities requires careful consideration of the complexity, clarity, and structure of the interface.

Clear and Simple Language

Using clear and simple language is essential for users with cognitive disabilities. Developers should avoid technical jargon and complex terminology, opting instead for plain language that is easy to understand. Instructions and explanations should be concise and supplemented with visual aids when necessary.

Consistent Layouts and Navigation

For users with cognitive disabilities, consistency is key. A consistent layout helps users understand how the software works and where to find specific features. Developers should ensure that the layout remains predictable across different screens and sections of the application.

Breaking Tasks into Manageable Steps

Complex tasks should be broken down into smaller, manageable steps to reduce cognitive load. For example, a multistep form can be divided into individual sections, with clear instructions and progress indicators guiding the user through each step.

Error Prevention and Recovery

Users with cognitive disabilities may struggle with error handling, so it's important to design software that minimizes the potential for mistakes. This includes providing confirmation dialogs for critical actions (such as deleting files) and offering undo or redo options to correct errors. When errors do occur, the system should provide clear instructions on how to resolve the issue.

Reducing Distractions

Users with attention disorders may find it difficult to focus on tasks if there are too many distractions on the screen. Developers should minimize unnecessary animations, flashing elements, or pop-ups that could interrupt the user's flow. Providing a distraction-free mode or simplified version of the interface can also improve accessibility for these users.

Mobile Accessibility Considerations

Mobile devices present unique challenges for accessibility due to their smaller screens, touch-based input, and varied usage environments. However, mobile platforms also offer built-in accessibility features that developers can leverage to create more inclusive applications.

Responsive and Scalable Interfaces

Mobile applications must be designed to adapt to a variety of screen sizes and orientations. A responsive design ensures that content remains readable and interactive, regardless of the device's screen dimensions. Text and interactive elements should be scalable, allowing users to increase the size of buttons, icons, and text without compromising functionality.

Touch Targets and Gestures

Interactive elements, such as buttons and links, should be large enough to be easily tapped by users with motor impairments. WCAG recommends a minimum touch target size of 44×44 pixels. Additionally, developers should provide alternatives to gesture-based interactions, such as swipe or pinch-to-zoom, for users who cannot perform complex touch gestures.

Voice Interaction

Mobile devices are often equipped with voice assistants, such as Siri or Google Assistant, that allow users to control their devices using voice commands. Developers can enhance mobile accessibility by integrating voice interaction into their apps, allowing users to navigate, search, or perform tasks using voice commands.

Haptic Feedback

Haptic feedback uses vibrations or tactile cues to provide feedback to users, offering a nonvisual way to confirm interactions. For users with visual impairments or those working in noisy environments, haptic feedback can indicate that a button has been pressed or an action has been completed.

Screen Reader Compatibility on Mobile

Mobile screen readers, such as VoiceOver (iOS) and TalkBack (Android), are essential for users with visual impairments. Developers should ensure that all interactive elements are labeled appropriately and that the application's layout is navigable using screen reader gestures. Testing the app with these screen readers can help identify and fix accessibility issues.

Web Accessibility Best Practices

Web accessibility ensures that websites and web applications are usable by people with disabilities, regardless of the device or platform they are using. Following web accessibility best practices helps developers create inclusive digital experiences that meet legal requirements and provide equal access to all users.

Semantic HTML

Using semantic HTML is essential for making web content accessible. Semantic HTML refers to using the correct HTML elements to structure content (e.g., <h1> for headings, <ul> for lists, and <button> for interactive elements). Proper use of semantic HTML ensures that assistive technologies, such as screen readers, can accurately interpret the content and convey it to users.

Alt Text for Images

Providing descriptive alternative text (alt text) for images is critical for users who rely on screen readers. Alt text should convey the meaning or purpose of the image, allowing users to understand its context even if they cannot see it. For decorative images that do not contribute to the content, the alt attribute should be left empty (alt="") to avoid cluttering the screen reader's output.

Accessible Forms

Forms are an essential part of many web applications, but they can present significant accessibility challenges if not designed correctly. To make forms accessible, developers should do the following.

- Use <label> elements to associate labels with form fields.

- Ensure that the required fields are clearly indicated.

- Provide clear error messages and instructions for correcting mistakes.

- Use ARIA attributes, such as `aria-required` or `aria-invalid`, to provide additional context for screen readers.

Keyboard Navigation

All interactive elements, such as buttons, links, and form fields, must be accessible via keyboard navigation. This means ensuring that users can navigate through the interface using the Tab key and that the focus state is clearly indicated. Developers should also provide keyboard shortcuts for common actions to improve usability for users who cannot use a mouse.

Accessible Media

Multimedia content, such as videos and audio, should be made accessible by providing captions, transcripts, and accessible media controls. **Captions** should accurately transcribe spoken dialogue and describe important non-verbal sounds. **Transcripts** should provide a readable version of audio content for users who are deaf or hard of hearing.

Color Contrast and Text Resizing

Text and background colors must have sufficient contrast to be readable by users with visual impairments. WCAG recommends a contrast ratio of at least 4.5:1 for normal text and 3:1 for large text. Additionally, web content should allow users to resize text without breaking the layout or functionality of the interface.

Avoiding Time Limits

Many users with disabilities may need more time to complete tasks, such as filling out forms or reading content. Developers should avoid imposing strict time limits on interactions or provide options for users to extend the time limit if necessary.

Accessibility Testing: Tools and Techniques

Accessibility testing is an essential step in ensuring that software meets the needs of users with disabilities. By identifying and addressing accessibility barriers early in the development process, developers can create more inclusive software that complies with legal requirements and enhances the user experience for all.

Types of Accessibility Testing

The following describes accessibility testing types.

- **Automated testing**: Automated accessibility testing tools scan the software for common accessibility issues, such as missing alt text, insufficient color contrast, or inaccessible form fields. While automated testing can identify many potential issues, it should not be relied on exclusively, as it may miss more complex problems that require manual testing.

- **Manual testing**: Manual testing involves simulating the experiences of users with disabilities, such as navigating the software using only a keyboard, testing with screen readers, or using alternative input devices. Manual testing is essential for identifying issues that

automated tools may not detect, such as the usability
of the interface for individuals with cognitive or motor
impairments.

- **Assistive technology testing**: Testing the software
with real assistive technologies, such as screen readers,
magnifiers, or voice recognition tools, provides valuable
insights into how the software performs for users with
disabilities. This type of testing helps ensure that the
software is compatible with the tools that users rely on.

- **User testing with people with disabilities**: Involving
users with disabilities in the testing process is one
of the most effective ways to identify accessibility
issues and gain insights into how the software can be
improved. User testing allows developers to observe
real-world interactions and understand the challenges
that users with disabilities face.

Popular Accessibility Testing Tools

- **WAVE (Web Accessibility Evaluation Tool)**: WAVE
is a browser extension that analyzes web pages for
accessibility issues and provides visual feedback
on potential problems. It highlights areas that need
improvement, such as missing alt text, low contrast, or
improperly labeled form fields.

- **Axe accessibility checker**: Axe is an open-source
accessibility testing tool that scans web pages for
accessibility issues. It provides detailed reports and
recommendations for fixing common problems, such
as missing ARIA attributes, keyboard navigation issues,
and form accessibility.

- **Lighthouse (Google Chrome)**: Lighthouse is a built-in tool in Google Chrome that performs automated testing for accessibility, performance, SEO, and other factors. It generates an accessibility score and provides recommendations for improving the accessibility of web pages.

- **Color contrast analyzers**: Tools like TPGi's Colour Contrast Analyser check the contrast between text and background colors to ensure they meet WCAG contrast ratio requirements.

- **Screen reader testing tools**: Screen readers like NVDA, JAWS, and VoiceOver (for macOS and iOS) can be used to test how software interacts with screen readers. These tools simulate the experience of users with visual impairments and help identify issues with screen reader compatibility.

- **Keyboard-only testing tools**: Developers can simulate keyboard-only navigation by using the Tab key to navigate through the interface. This helps identify areas where keyboard accessibility may be lacking, such as missing focus indicators or inaccessible buttons.

Best Practices for Accessibility Testing

The following describes some best practices for accessibility testing.

- **Combine automated and manual testing.** While automated tools can identify many common accessibility issues, they cannot catch everything. Manual testing, particularly with real assistive technologies and users with disabilities, is essential for ensuring comprehensive accessibility.

- **Test early and often.** Accessibility testing should be conducted throughout the development process, not just at the end. By testing early and often, developers can identify and address issues before they become more difficult and costly to fix.

- **Use real devices and assistive technologies.** Testing on real devices (e.g., mobile phones, tablets, desktop computers) and with real assistive technologies (e.g., screen readers, magnifiers) provides the most accurate insights into how users with disabilities will interact with the software.

By incorporating accessibility testing into the development process, developers can create software that is more inclusive and accessible to users with disabilities, ensuring a better overall user experience.

Real-World Examples of Accessibility in Action

Several leading technology companies and organizations have implemented innovative accessibility features in their software, demonstrating how accessibility can be integrated into mainstream applications to create more inclusive experiences.

Example 1: Apple's VoiceOver

Apple's VoiceOver is a built-in screen reader available on macOS, iOS, and iPadOS devices. VoiceOver allows users with visual impairments to navigate and interact with their devices using auditory feedback. Users can swipe, tap, or use voice commands to control their device, while VoiceOver describes what is happening on the screen. VoiceOver supports a wide range of gestures, allowing users to perform actions such as selecting items, scrolling through lists, or activating buttons. It also integrates with braille displays, providing tactile feedback for users who are blind.

Apple's commitment to accessibility is evident in its focus on providing high-quality, built-in assistive technologies, ensuring that users with disabilities can access and interact with their devices without the need for additional software.

Example 2: Microsoft's Accessibility Features

Microsoft has integrated a wide range of accessibility features into its Windows operating system, making it easier for users with disabilities to interact with their computers. The following are some key accessibility features.

- **Narrator**: A built-in screen reader that provides spoken feedback to users with visual impairments.

- **Magnifier**: A screen magnification tool that enlarges text, images, and other on-screen elements for users with low vision.

- **Speech Recognition**: A tool that allows users to control their computer and input text using voice commands, providing an alternative to traditional keyboard and mouse input.

- **High-Contrast Mode**: A display setting that increases the contrast between text and background colors, making it easier for users with visual impairments to read on-screen content.

Microsoft also offers comprehensive support for third-party assistive technologies, such as screen readers, alternative input devices, and eye-tracking systems. Through its ongoing commitment to accessibility, Microsoft ensures that its products are usable by individuals with a wide range of disabilities.

Example 3: Facebook's AI-Powered Alt Text

Facebook has implemented an AI-powered system that automatically generates alt text for images, helping users with visual impairments understand the content of photos on the platform. The system uses machine learning to analyze images and generate a brief description of the objects and people in the photo. While the descriptions are not always perfect, they provide valuable context for users who rely on screen readers to navigate the platform.

This feature demonstrates how AI can be leveraged to enhance accessibility by automatically generating content that would otherwise require manual input. By making visual content more accessible, Facebook ensures that users with visual impairments can engage more fully with the platform.

Example 4: Google's Live Transcribe

Google's Live Transcribe is an Android app that provides real-time transcription of speech, making it easier for users who are deaf or hard of hearing to participate in conversations. The app uses Google's speech recognition technology to transcribe spoken words into text, which is displayed on the screen. Live Transcribe supports multiple languages and dialects, and users can adjust the text size and color contrast for better readability.

By offering an easy-to-use, real-time transcription tool, Google enables users with hearing impairments to engage in conversations more easily, whether in-person or over the phone.

These real-world examples highlight how leading technology companies are integrating accessibility features into their products to create more inclusive digital experiences. By leveraging assistive technologies and innovative design solutions, these companies are setting a standard for accessible software development.

Accessibility and Legal Compliance

Ensuring that software is accessible is not only a matter of inclusivity but also a legal requirement in many regions. Failing to meet accessibility standards can result in lawsuits, fines, and reputational damage. Understanding the legal landscape surrounding accessibility is essential for organizations that want to avoid legal issues and create compliant software.

Key Accessibility Laws and Regulations

- **Americans with Disabilities Act (ADA) – United States**: The ADA is a civil rights law that prohibits discrimination against individuals with disabilities. While the ADA does not explicitly address digital accessibility, courts have increasingly interpreted its provisions to apply to websites and digital services, requiring businesses to make their websites accessible to individuals with disabilities.

- **Section 508 of the Rehabilitation Act – United States**: Section 508 requires federal agencies to make their electronic and information technology accessible to individuals with disabilities. This includes websites, software applications, documents, and multimedia. Section 508 compliance is mandatory for federal agencies and contractors, and it closely aligns with WCAG 2.0 standards.

- **European Accessibility Act – European Union**: The European Accessibility Act mandates that certain products and services, including websites, mobile applications, and e-commerce platforms, must be accessible to individuals with disabilities. The Act requires conformance with WCAG 2.1 standards and applies to public sector bodies, businesses, and other organizations.

- **UK Equality Act 2010 – United Kingdom**: The Equality Act requires businesses and organizations to make reasonable adjustments to ensure that individuals with disabilities can access their services. This includes making websites and digital services accessible. Failure to comply with the Equality Act can result in legal action and fines.

- **AODA (Accessibility for Ontarians with Disabilities Act) – Canada**: The AODA requires public and private organizations in Ontario to make their websites and digital content accessible to individuals with disabilities. The AODA is based on WCAG 2.0 Level AA standards and applies to businesses, non-profits, and government agencies.

Consequences of Non-Compliance

- **Lawsuits and legal action**: Organizations that fail to comply with accessibility laws may face lawsuits from individuals or advocacy groups. In the United States, there has been a significant increase in lawsuits related to website accessibility under the ADA. These lawsuits can result in costly settlements and damage to the organization's reputation.

- **Fines and penalties**: Governments may impose fines or penalties on organizations that fail to meet accessibility standards. For example, under the AODA in Ontario, organizations can face fines of up to $100,000 per day for non-compliance.

- **Loss of business**: Inaccessible websites and software can result in lost business opportunities, as individuals with disabilities may be unable to access the organization's products or services. This can lead to a decrease in customer satisfaction and damage to the organization's brand.

Ensuring Legal Compliance

Organizations should do the following to ensure that software is legally compliant with accessibility standards.

- **Adopt WCAG standards.** WCAG 2.0 and 2.1 standards are widely recognized as the benchmark for digital accessibility. By adhering to these standards, organizations can ensure that their software meets legal requirements in most regions.

- **Conduct regular accessibility audits.** Regularly auditing software for accessibility compliance can help identify and address issues before they result in legal action. Audits should include both automated and manual testing to ensure comprehensive coverage.

- **Provide training for developers.** Ensuring that developers are trained in accessibility best practices is essential for creating compliant software. Training should cover topics such as semantic HTML, ARIA attributes, and assistive technology compatibility.

- **Involve legal experts.** Consulting with legal experts who specialize in accessibility law can help organizations navigate the complex legal landscape and ensure that their software is fully compliant with relevant regulations.

By understanding the legal requirements for accessibility and taking proactive steps to ensure compliance, organizations can avoid legal challenges and create software that is accessible to all users.

The Role of Assistive Technology in Enhancing Accessibility

Assistive technologies play a critical role in helping individuals with disabilities interact with software and digital content. These technologies bridge the gap between inaccessible content and users' needs, providing alternative ways to access, navigate, and interact with software. Understanding the role of assistive technology is essential for developers who want to create accessible software that supports a wide range of users.

Types of Assistive Technology

- **Screen readers**: Screen readers convert on-screen text into speech or braille output, allowing users with visual impairments to navigate and interact with software. Popular screen readers include NVDA, JAWS, and VoiceOver. To ensure compatibility with screen readers, developers must use semantic HTML, provide descriptive alt text for images, and ensure that interactive elements are labeled correctly.

- **Magnification tools**: Magnification tools enlarge text, images, and other on-screen elements to make them more readable for users with low vision. Built-in tools like Microsoft's Magnifier or macOS's Zoom allow users to zoom in on specific areas of the screen. Developers should ensure that their software supports magnification by allowing text to resize without breaking the layout or functionality of the interface.

- **Alternative input devices**: Alternative input devices, such as adaptive keyboards, switch systems, and eye-tracking devices, allow users with motor impairments to interact with software. For example, a user with limited mobility may use a switch system to control their computer by pressing a single button or an eye-tracking device to move the cursor and select items on the screen. Ensuring that the software is compatible with these devices is essential for accessibility.

- **Voice recognition software**: Voice recognition software, such as Dragon NaturallySpeaking or built-in voice assistants like Google Assistant and Siri, allows users to control their devices and input text using voice commands. This technology is particularly useful for users with motor impairments who may have difficulty using a traditional keyboard or mouse. Developers can integrate voice recognition capabilities into their software by using APIs like Google's Speech-to-Text or Apple's Voice Control.

- **Braille displays**: Braille displays convert on-screen text into braille output for users who are blind. These devices allow users to read and navigate digital content using tactile feedback. Developers should ensure that their software is compatible with braille displays by providing accessible text alternatives for all visual content and supporting screen reader technologies that work with braille displays.

- **Assistive touch and gesture control**: Assistive touch features, such as those available on iOS devices, allow users with motor impairments to perform gestures and interact with touch screens using alternative input methods. Developers should ensure that their software is compatible with assistive touch features and provide alternatives to gestures that may be difficult for users with limited dexterity.

Enhancing Accessibility with Assistive Technology

- **Testing with assistive technology**: To ensure that their software is compatible with assistive technologies, developers should test their software using real assistive devices and tools. This includes testing with screen readers, magnifiers, voice recognition software, and alternative input devices to identify and address any compatibility issues.

- **Providing customizable interfaces**: Users who rely on assistive technology may need to customize the interface to meet their specific needs. Providing options to adjust text size, contrast levels, input sensitivity, and other settings can make the software more accessible and adaptable to different assistive devices.

- **Supporting ARIA (Accessible Rich Internet Applications)**: ARIA attributes provide additional context for assistive technologies, helping them interpret and convey the meaning of interactive elements such as buttons, links, and forms. By using ARIA attributes correctly, developers can enhance the accessibility of their software for users who rely on assistive technology.

Assistive technology is a powerful tool for enhancing accessibility, enabling individuals with disabilities to interact with software in ways that suit their abilities. By designing software that is compatible with a wide range of assistive technologies, developers can create more inclusive digital experiences that meet the needs of all users.

Designing for Cultural and Linguistic Accessibility

Cultural and linguistic accessibility focuses on making software usable by individuals from diverse cultural, linguistic, and geographic backgrounds. This includes accommodating different languages, writing systems, and cultural norms, as well as ensuring that software is inclusive and respectful of diverse user experiences.

Key Considerations for Cultural and Linguistic Accessibility

- **Multilingual support**: To make software accessible to users who speak different languages, developers should provide multilingual support. This involves offering the software interface, documentation, and help content in multiple languages, as well as ensuring that users can easily switch between languages. Localization is the process of adapting the software to different languages and cultures, including translating text and adjusting formatting for dates, times, and currencies.

- **Right-to-Left (RTL) language support**: Some languages, such as Arabic, Hebrew, and Persian, are written from right to left. Ensuring that the software supports RTL languages involves adjusting the layout, text alignment, and navigation patterns to accommodate RTL reading and writing. Developers should test their software in RTL languages to ensure that all elements are properly aligned and displayed.

- **Cultural sensitivity**: Developers must be aware of cultural differences when designing software, including differences in symbols, colors, gestures, and icons. For example, a hand gesture that is acceptable in one culture may be offensive in another, or a color that is considered lucky in one culture may have negative connotations in another. Ensuring that the software is culturally sensitive and inclusive can prevent misunderstandings and enhance the user experience.

- **Inclusive language**: The language used in the software should be inclusive and respectful of all users. This includes avoiding gendered language, using neutral terms when referring to individuals, and being mindful of cultural differences in language use. Providing alternative language options, such as nonbinary pronouns or inclusive terms for different gender identities, can make the software more inclusive for users from diverse backgrounds.

79

- **Supporting multiple writing systems**: Different languages use different writing systems, such as the Latin alphabet, Cyrillic script, or Chinese characters. Developers should ensure that their software supports a wide range of writing systems and that text is displayed correctly regardless of the language or script being used. This includes supporting non-Latin characters, special diacritics, and combining characters where necessary.

- **Accessibility for non-native speakers**: Non-native speakers may find it more challenging to navigate and understand software that is written in a language they are not fluent in. To make the software more accessible to non-native speakers, developers should use simple, clear language and avoid jargon or complex terms. Providing visual aids, such as icons or diagrams, can also help users understand the content more easily.

Tools for Testing Cultural and Linguistic Accessibility

- **Translation management system (TMS)**: Tools like Smartling, Crowdin, or Phrase can help developers manage the translation and localization process, ensuring that the software is accurately translated into multiple languages.

- **Language and locale testing**: Developers should test their software in different languages and locales to ensure that all text is properly translated and formatted. This includes testing for RTL language support, character encoding issues, and cultural appropriateness of icons and symbols.

- **Inclusive language checkers**: Tools like Microsoft's Inclusive Language Checker can help developers identify and replace non-inclusive language in their software, ensuring that the language is respectful and inclusive of all users.

By designing for cultural and linguistic accessibility, developers can create software that is inclusive and accessible to users from diverse backgrounds, ensuring that language and cultural differences do not become barriers to usability.

Continuous Improvement in Accessibility

Accessibility is not a one-time effort but an ongoing process of improvement. As technologies evolve and user needs change, software must continually adapt to meet new accessibility challenges. By adopting a mindset of continuous improvement, developers can ensure that their software remains accessible and usable for all users over time.

Key Strategies for Continuous Improvement in Accessibility

- **Accessibility audits**: Regularly conducting accessibility audits helps identify areas where the software may be falling short of accessibility standards. Audits should include both automated testing and manual reviews, as well as testing with real users who have disabilities. By regularly auditing the software, developers can ensure that it remains compliant with accessibility standards and continues to meet the needs of users with disabilities.

- **User feedback**: Gathering feedback from users with disabilities is one of the most effective ways to identify accessibility issues and opportunities for improvement. Developers should encourage users to report accessibility barriers they encounter and take their feedback into account when planning updates and improvements.

- **Staying up-to-date with accessibility standards**: Accessibility standards, such as WCAG, are periodically updated to reflect new technologies and emerging best practices. Developers should stay informed about updates to accessibility standards and ensure that their software remains compliant with the latest guidelines.

- **Ongoing training for developers**: Ensuring that developers are continuously trained in accessibility best practices is essential for maintaining accessible software. This includes keeping developers informed about new assistive technologies, accessibility tools, and legal requirements. Ongoing training helps developers stay current with accessibility trends and ensures that they have the skills needed to create inclusive software.

- **Inclusive design from the start**: Accessibility should be considered from the earliest stages of the design and development process rather than as an afterthought. By incorporating accessibility into the design process from the beginning, developers can create software that is more inclusive and reduces the need for retroactive fixes.

- **Agile accessibility**: Agile development processes, which emphasize iterative development and continuous feedback, provide an ideal framework for ongoing accessibility improvement. By integrating accessibility testing and feedback into each development cycle, developers can identify and address accessibility issues more quickly and ensure that improvements are made continuously.

- **Adapting to new technologies**: As new technologies, such as artificial intelligence, augmented reality, and virtual reality, become more prevalent, developers must ensure that these technologies are accessible to users with disabilities. This may involve developing new accessibility guidelines or integrating emerging assistive technologies to support users in these new environments.

By committing to continuous improvement in accessibility, developers can ensure that their software remains inclusive and responsive to the changing needs of users with disabilities, creating a better overall user experience for all.

The Future of Accessibility and Emerging Technologies

As technology continues to evolve, new challenges and opportunities for accessibility will emerge. Innovations such as artificial intelligence (AI), augmented reality (AR), virtual reality (VR), and the Internet of Things (IoT) are reshaping the digital landscape, and ensuring that these technologies are accessible is essential for creating an inclusive future.

Artificial Intelligence and Accessibility

AI has the potential to transform accessibility by automating tasks and providing personalized assistance to users with disabilities. The following are examples.

- **AI-powered transcription services** can provide real-time captions for video and audio content, making media more accessible to users who are deaf or hard of hearing.

- **AI-driven screen readers** can enhance the experience for users with visual impairments by providing more accurate and context-aware descriptions of on-screen elements.

- **Natural language processing (NLP)** can help users with cognitive disabilities by simplifying complex text and providing alternative explanations for difficult concepts.

However, developers must ensure that AI systems are trained to recognize and accommodate the needs of users with disabilities, avoiding bias and ensuring that AI-powered tools are inclusive.

Augmented Reality and Virtual Reality

AR and VR technologies are becoming increasingly popular in gaming, education, and professional training. However, these immersive environments present unique accessibility challenges, particularly for users with sensory or mobility impairments.

To make AR and VR experiences accessible, developers should do the following.

- Provide **alternative input methods** for users who cannot perform physical gestures or use traditional controllers.

- Offer **audio descriptions** for visual elements in AR and VR environments, allowing users with visual impairments to engage with the content.

- Ensure that **haptic feedback** and other sensory cues are available for users with hearing impairments.

As AR and VR technologies evolve, creating accessible, immersive experiences will become increasingly important in ensuring that all users can participate in these digital worlds.

The Internet of Things and Accessibility

IoT is revolutionizing the way people interact with everyday objects, from smart home devices to wearable technology. For users with disabilities, IoT devices offer new opportunities for independence and convenience. However, accessibility must be a priority in the design of IoT systems.

To ensure that IoT devices are accessible, developers should do the following.

- Provide **voice control** and **alternative input methods** for users with mobility impairments.

- Ensure that **IoT interfaces** are compatible with screen readers and other assistive technologies.

- Offer **customizable settings** that allow users to adjust device behavior based on their needs and preferences.

As IoT becomes more integrated into daily life, accessible design will be key to ensuring that people with disabilities can fully benefit from these innovations.

Ethical Considerations in Emerging Technologies

As new technologies are developed, ethical considerations surrounding accessibility must be addressed. Developers must ensure that emerging technologies do not exacerbate existing inequalities or create new barriers for users with disabilities. This includes the following.

- **Preventing bias** in AI systems that could disadvantage users with disabilities.

- **Ensuring privacy** for users of assistive technologies, particularly when it comes to sensitive health or personal information.

- **Fostering inclusion** by involving users with disabilities in the design and development of new technologies.

By taking a proactive approach to accessibility in emerging technologies, developers can create a future where innovation benefits everyone, regardless of their abilities.

Conclusion

Enhancing software accessibility is essential for creating inclusive digital experiences that cater to the diverse needs of users with disabilities. By understanding and addressing the specific accessibility requirements of individuals with visual, auditory, motor, cognitive, and linguistic disabilities, developers can remove barriers to access and create software that is usable by all.

This chapter explored the principles, guidelines, and best practices that underpin accessible software design, from universal design principles to the role of assistive technologies and cultural accessibility. It also highlighted the importance of legal compliance, continuous improvement, and user feedback in maintaining and improving accessibility over time.

Ultimately, accessibility is about more than just meeting legal requirements—it is about ensuring that everyone, regardless of their abilities, can participate fully in the digital world. By prioritizing accessibility in the design and development process, developers can create software that is not only functional but also empowering, fostering greater inclusion and equal access for all users.

CHAPTER 4

Data Standardization and Interoperability

In the realm of software and data management, the concepts of data standardization and interoperability stand out as fundamental pillars. They ensure that data, regardless of its source, can be effectively shared, understood, and utilized across various systems and platforms. This chapter delves into the definitions, significance, and practical applications of these concepts, highlighting their critical role in fostering innovation, efficiency, and collaboration. By establishing common formats, structures, and protocols, data standardization paves the way for seamless data exchange and integration. Meanwhile, interoperability empowers different software systems and applications to communicate and cooperate effectively, breaking down information silos and enabling a more connected digital landscape. Both concepts play a pivotal role in unlocking the full potential of data, empowering organizations to make informed decisions, streamline processes, and drive transformative change across industries.

What Is Data Standardization?

Data standardization refers to the process of converting data from various sources into a common format, making it consistent and comparable. This involves aligning data types, formats, and values to adhere to predefined norms and standards.

The Importance of Data Standardization

The profound significance of data standardization reaches far beyond the realm of mere data management practices. It emerges as a cornerstone for organizations that aspire to unlock the full potential of their data assets. This expanded exploration delves deeper into the multifaceted advantages of data standardization, encompassing the pillars of consistency, efficiency, and comparability.

Consistency

Data standardization is pivotal in achieving data consistency across various systems and platforms within an organization. In today's data-driven world, where data is collected from a myriad of sources, including IoT devices, online transactions, and user interactions, the risk of discrepancies and incompatibilities is high. Standardization processes, such as normalization of formats, alignment of data types, and adoption of common data models, ensure that all data, regardless of its origin, adhere to a uniform structure and format. This consistency is crucial for the following.

- **Accurate analysis**: Analytical models and algorithms depend on reliable data to generate meaningful insights. Consistent data reduces the noise and variability that can lead to inaccurate analyses and faulty conclusions. When data is standardized, it's easier to identify patterns, trends, and correlations, leading to more robust and trustworthy analytical outcomes.

- **Informed decision-making**: Organizations rely on data to make strategic decisions. Consistent data provides a trustworthy foundation for decision-makers, enabling them to base their strategies on solid, dependable information. Decisions made on standardized data are less likely to be skewed by inconsistencies or errors, leading to better outcomes and improved organizational performance.

Efficiency

Efficiency in data management is another significant benefit of data standardization. Handling data from diverse sources often involves manual adjustments and conversions to ensure compatibility, which is both time-consuming and prone to errors. Standardization automates many of these processes, streamlining data ingestion, storage, and retrieval. This efficiency manifests in several ways.

- **Reduced errors**: Automated processes are less susceptible to the errors that commonly occur during manual data handling, leading to cleaner, more reliable datasets. By minimizing human intervention, standardization reduces the risk of introducing errors during data manipulation, resulting in higher data quality.

- **Time savings**: Standardization reduces the need for repetitive manual data cleaning and formatting tasks, freeing up valuable time for data analysts and scientists to focus on higher-value activities such as data analysis and interpretation. The time saved through automation can be redirected toward more strategic tasks, accelerating the overall data analysis lifecycle.

- **Resource optimization**: By simplifying data management workflows, organizations can allocate their resources more effectively, investing in innovation and strategic initiatives rather than routine data maintenance. This optimization allows organizations to maximize the value derived from their data assets, driving growth and competitiveness.

Comparability

Comparability is essential for conducting meaningful analytics and research, and data standardization lays the groundwork for comparable data. It enables the aggregation, analysis, and comparison of data from varied sources, facilitating cross-system and cross-organizational studies that would otherwise be fraught with compatibility issues. This comparability is vital for the following.

- **Benchmarking and performance analysis**: Standardized data allows organizations to benchmark their performance against industry standards or competitors by ensuring that the datasets being compared are compatible. This comparison provides valuable insights into areas for improvement and helps organizations identify best practices.

- **Collaborative research and innovation**: In fields such as healthcare, environmental science, and public policy, data standardization enables researchers and practitioners from different organizations and disciplines to collaborate more effectively, combining their data for comprehensive analyses. Standardized data fosters a common language for data sharing, accelerating research and driving innovation across diverse fields.

- **Regulatory compliance**: Many industries are subject to regulatory requirements mandating the reporting and analysis of data in standardized formats. Compliance is streamlined when data is standardized, reducing the risk of regulatory breaches and associated penalties. By adhering to standardized data formats, organizations can ensure they meet regulatory obligations while minimizing the effort and resources required for compliance.

What Is Interoperability?

While standardization is critical for consistent and reliable data, it also plays a foundational role in achieving interoperability across systems and platforms. Interoperability is the capability of diverse information systems, devices, or applications to seamlessly access, exchange, integrate, and utilize data in a coordinated manner. This collaborative functionality extends both within and across organizational boundaries, ensuring smooth data flow and usability.

Levels of Interoperability

Interoperability is a multifaceted concept crucial for effective communication and data exchange across different systems, platforms, or organizations. It's comprised of several layers, each with distinct requirements that contribute to the overall goal of seamless data sharing and utilization. Understanding these levels is key for organizations seeking to maximize their interoperability capabilities.

Foundational Interoperability

This foundational level establishes the core technical prerequisites for any data exchange. It focuses on enabling systems and applications to connect and communicate without necessarily requiring the receiving system to fully interpret the data.

- **Technical connectivity**: Setting up the necessary infrastructure (networks, protocols, interfaces) to ensure data can be reliably transmitted and received across systems.

- **Data exchange protocols**: Employing standardized protocols to define how data is transferred, ensuring compatibility among diverse platforms.

- **Security measures**: Implementing security protocols to safeguard data integrity and confidentiality during transmission.

Structural Interoperability

Building upon the foundational level, structural interoperability defines the format, syntax, and organization of exchanged data. This ensures that receiving systems can automatically process and understand the data's structure.

- **Data format standards**: Utilizing standardized formats like XML or JSON to specify how information is encoded and arranged, making it easier for systems to parse and use the data.

- **Schema and syntax rules**: Establishing schemas and rules that govern the organization of data elements and their relationships, enabling automated data processing.

- **Message standards**: Employing standards that define the layout and sequence of information within a data exchange, ensuring correct interpretation by receiving systems.

Semantic Interoperability

The most sophisticated level, semantic interoperability, focuses on preserving the meaning and context of exchanged information. Both sending and receiving systems understand not just the structure but also the intended interpretation of the data.

- **Common data models**: Utilizing shared models and ontologies to provide a common understanding of concepts and terms within the data, bridging semantic gaps between systems.

- **Contextual data use**: Enabling data to be used in diverse contexts while retaining its original meaning and relevance.

- **Actionable information**: Ensuring that received data is not only interpretable but also actionable, allowing systems to make informed decisions based on the information.

Each level of interoperability builds upon and complements the previous one, forming a comprehensive framework for efficient, meaningful, and secure data exchange. By embracing these levels, organizations can significantly enhance their capacity to collaborate, innovate, and deliver services effectively in the interconnected digital landscape.

Standards and Frameworks

International standards and frameworks are essential in steering data standardization and interoperability across various industries and domains. By adhering to these guidelines, organizations can ensure their data management practices are robust, secure, and efficient, facilitating seamless communication and data exchange with other systems and entities. Let's delve deeper into some notable standards and frameworks that have been instrumental in shaping the landscape of data standardization and interoperability.

ISO/IEC Standards

The International Organization for Standardization (ISO) and the International Electrotechnical Commission (IEC) jointly offer a comprehensive suite of standards that provide in-depth guidelines on data management, security, and quality. These standards cover a wide spectrum of topics, including information security management (ISO/IEC 27001), data quality (ISO 8000), and cloud computing (ISO/IEC 17788).

- **Guidelines on data management**: ISO/IEC standards offer detailed best practices on how to meticulously organize, store, and maintain data to ensure its accuracy, completeness, and reliability throughout its lifecycle.

- **Security measures**: They provide comprehensive frameworks for establishing, implementing, maintaining, and continuously improving information security management systems, encompassing technical and organizational measures to safeguard data against unauthorized access, breaches, and other threats.

- **Quality assurance**: These standards also define a range of metrics and processes for ensuring that data meets specific quality benchmarks, which is fundamental for trustworthy data analysis, informed decision-making, and maintaining the integrity of data-driven processes.

HL7 and FHIR for Healthcare

Health Level Seven International (HL7) and Fast Healthcare Interoperability Resources (FHIR) are specialized standards tailored for the healthcare industry, focusing on the exchange, integration, sharing, and retrieval of electronic health information in a standardized and interoperable manner.

- **HL7 standards** enable the seamless exchange of clinical and administrative data among disparate hospital information systems, electronic health records, and other healthcare applications. These standards ensure that different systems can communicate effectively, promoting better coordination of care, reducing errors, and ultimately improving patient outcomes.

- **FHIR** is a modern standard that leverages the latest web standards and a RESTful API approach to streamline the exchange of healthcare information. It supports a wide array of applications, including mobile apps, cloud communications, and data analytics tools, enhancing interoperability in the healthcare ecosystem and fostering innovation in health IT solutions.

EDIFACT for Commerce

The Electronic Data Interchange for Administration, Commerce, and Transport (EDIFACT) is an international standard developed by the United Nations. It specifies a detailed set of syntax rules for electronic data interchange between independent computer systems, fostering seamless communication and data exchange in global commerce.

- **Global commerce:** EDIFACT is widely adopted in international trade, enabling businesses to exchange a variety of documents such as purchase orders, invoices, shipping notices, and customs declarations in a standardized electronic format. This facilitates efficient cross-border transactions and reduces the reliance on paper-based processes.

- **Efficiency and accuracy:** By standardizing the format and content of these documents, EDIFACT significantly reduces the potential for errors and misinterpretations that can arise from manual data entry or incompatible systems. This streamlines business processes, improves operational efficiency, and minimizes costs associated with data discrepancies.

- **Cross-industry applicability:** While particularly prevalent in commerce, EDIFACT's principles and structure can be adapted to various other industries, including healthcare, transportation, and finance, promoting broad-based interoperability and data exchange across diverse sectors.

These standards and frameworks are essential for harmonizing data practices across borders and sectors, fostering seamless data sharing and interoperability on a global scale. By implementing these guidelines, organizations can enhance their ability to engage in secure and efficient data exchange, promoting collaboration and driving innovation in an increasingly interconnected world.

Challenges in Implementation

Operationalizing data standardization and interoperability across diverse systems and platforms presents a multitude of challenges. These challenges can be formidable for organizations seeking to harness the full potential of seamless data exchange. Understanding these obstacles in detail is the first step toward devising effective strategies to overcome them.

Complexity

The integration of heterogeneous systems and diverse data formats is inherently complex. Organizations often utilize a wide array of software and hardware solutions from different vendors, each with its own unique data formats, protocols, and idiosyncrasies.

- **System diversity**: The heterogeneity of IT systems and applications, ranging from legacy systems with outdated technologies to modern cloud-based solutions with cutting-edge features, presents a significant integration challenge. Bridging the gap between these disparate systems requires careful planning, expertise, and specialized tools.

- **Resource intensiveness**: The process of achieving interoperability is often resource-intensive, demanding significant investments of time, expertise, and financial resources. Aligning different systems, training personnel on new standards and protocols, and customizing solutions to ensure compatibility can strain organizational resources and require ongoing maintenance.

- **Continuous evolution**: The rapid pace of technological advancements and the continuous evolution of systems and standards mean that interoperability solutions must be adaptable and scalable to accommodate future changes. This requires a proactive approach to technology adoption and a commitment to ongoing learning and adaptation.

Data Privacy and Security

In an era marked by increasing data breaches and cyber threats, ensuring the privacy and security of data during exchange is of paramount importance. This challenge is amplified by the varied nature of data protection laws across jurisdictions, creating a complex regulatory landscape that organizations must navigate.

- **Vulnerable data transfers**: As data traverses between systems, it becomes vulnerable to interception, unauthorized access, and other security threats. Implementing robust encryption algorithms, secure data transfer protocols, and access controls is essential but can be technically challenging and resource-intensive.

- **Regulatory compliance**: Organizations must comply with a myriad of data protection regulations, such as the General Data Protection Regulation (GDPR) in Europe, the Health Insurance Portability and Accountability Act (HIPAA) in the US for healthcare information, and many others. Compliance requires in-depth knowledge of the laws, continuous monitoring of data handling practices, and adaptation to evolving legal frameworks.

- **Trust and reliability**: Building trust among stakeholders that data is handled securely and responsibly is crucial. This involves not only implementing technical solutions but also establishing transparent policies and practices regarding data use, sharing, and storage. Maintaining open communication with stakeholders and demonstrating a commitment to data protection can help build trust and foster confidence in data exchange processes.

Compliance

The need to adhere to a wide array of regulatory, legal, and industry-specific standards introduces a substantial layer of complexity to data standardization and interoperability efforts. Organizations must not only understand and comply with these standards but also adapt to their evolving nature.

- **Diverse regulations**: Different countries and industries have their own sets of regulations governing data handling, privacy, and exchange. Organizations operating across borders or in multiple sectors must navigate this complex regulatory landscape, ensuring compliance with all relevant laws and standards.

- **Evolving standards**: Legal and regulatory frameworks are constantly evolving to keep pace with technological advancements and emerging threats. Organizations

must remain vigilant and proactive in updating their
compliance strategies to align with these changes,
requiring ongoing monitoring and adaptation.

- **Penalties and reputational risk**: Non-compliance
  with data protection and privacy regulations can result
  in severe consequences, including hefty fines, legal
  action, and irreparable damage to an organization's
  reputation. Ensuring compliance is not just a legal
  obligation but a critical component of risk management
  and maintaining stakeholder trust.

Overcoming these multifaceted challenges requires a strategic and
holistic approach that encompasses investing in the right technologies,
fostering a culture of security and compliance throughout the
organization, and engaging in continuous learning and adaptation
to stay ahead of the evolving technological and regulatory landscape.
Collaboration between industry players, regulatory bodies, and technology
providers can also pave the way for more standardized, secure, and
interoperable data exchange practices in the future.

Case Studies

In the era of digital transformation, the healthcare sector is not left behind.
With the adoption of advanced technologies, healthcare administration
and management have seen significant improvements in service delivery.
This wave of digital transformation in healthcare is reshaping patient care,
powered by data standardization and interoperability. This represents
an annual growth rate of 36% and is growing faster than data from other
industries such as manufacturing, financial services, and media and
entertainment (see www.lek.com/insights/hea/eu/ei/tapping-new-
potential-realising-value-data-healthcare-sector). However, with

these advancements come challenges, one of which is the need for data standardization and interoperability. Let's delve into its crucial roles, which are important in managing this surge efficiently and effectively.

Image Source: DALL·E 2024-03-27 23.24.01

The Significance of Data Standardization and Interoperability in Healthcare

Data standardization is the process of making data fit a common format. This aids in collaborative research, large-scale data analysis, and the sharing of sophisticated tools and techniques. On the other hand, interoperability is when different systems, devices, and apps can work together to use data in a well-organized way across different organizations and regions.

Data standardization and interoperability hold a significant place in the healthcare sector. They simplify the process of data sharing among various healthcare providers, creating a more integrated and efficient system. This seamless exchange of information is crucial in providing timely and accurate patient care. For instance, a doctor can quickly access a patient's medical history from another clinic, leading to a more informed and effective treatment plan. Furthermore, these practices help consolidate different healthcare services. By enabling various systems to work together, redundancies are reduced, operations are streamlined, and costs are cut down. In essence, data standardization and interoperability are key drivers in enhancing the efficiency and cost-effectiveness of healthcare services.

Innovative Approaches and Best Practices

In healthcare IT, the emphasis on innovative approaches and best practices is always a focal point, particularly in the context of user-centric design and enhanced interoperability.

A noteworthy example of such innovation is the Trusted Exchange Framework and Common Agreement, or TEFCA, (see `https://telehealthandmedicinetoday.com/index.php/journal/article/view/428`). The TEFCA framework represents a significant advancement in the field, as it provides a standardized method for healthcare information networks to connect and share patient data securely and efficiently. One of the key findings in their study was a remarkable 40% reduction in medical errors, underscoring the critical role of efficient data sharing in enhancing the accuracy of diagnoses and treatment plans. This reduction in errors is attributed to the seamless exchange of patient information. It is instrumental in bridging the gap between different health information networks, thereby facilitating smoother and more reliable data exchange. The adoption of TEFCA has the potential to streamline patient care coordination, enhance clinical outcomes, and support public health reporting and analysis.

Another critical aspect of innovation in healthcare IT is the standardization of health terminologies, a subject thoroughly investigated by Sitompul et al (see `https://ieeexplore.ieee.org/document/10278050`). Their research highlights the necessity of adopting standardized medical terminologies across healthcare systems. Such standardization ensures consistency and uniformity in medical communication, which is vital for accurate and effective decision-making in healthcare. Standardized terminologies, like the International Classification of Diseases (ICD) and the Systematized Nomenclature of Medicine—Clinical Terms (SNOMED-CT), aid in reducing misunderstandings and errors in patient care. Standardized health terminologies play a pivotal role, as demonstrated in this study, where a systematic literature review reveals improved decision-making efficiency in healthcare by 25% through standardized terminologies. They also play a crucial role in healthcare analytics, where consistent terminology is essential for accurate data aggregation and analysis.

These innovative approaches, including frameworks like TEFCA and the standardization of health terminologies, are not just technical enhancements; they represent a paradigm shift toward a more interconnected and unified healthcare system. These advancements support the broader goals of healthcare IT, which include improving patient outcomes, enhancing the efficiency of healthcare delivery, and reducing the overall costs associated with healthcare provision. As the healthcare industry continues to evolve, the implementation of these best practices is expected to drive significant improvements in both patient care and healthcare management.

Real-Life Applications of Data Standardization and Interoperability in Healthcare

Electronic Health Records

Electronic health records, or EHRs, have brought about a significant change in the healthcare sector. They have made it possible for patient information to be shared among various healthcare providers. A survey in 2021 found that a large majority of doctors working in offices (78%) and almost all non-government hospitals (96%) have adopted a certified EHR system. This was a huge leap from 2011, when only 28% of hospitals and 34% of doctors had adopted an EHR system (see www.healthit.gov/data/ quickstats). The adoption of EHRs has led to improvements in patient care and has also helped in reducing costs. For example, EHRs can prevent the repetition of tests by giving healthcare providers access to a patient's past test results.

FHIR

FHIR is a standard set by Health Level Seven International (HL7) for the electronic exchange of healthcare information. The healthcare community is increasingly adopting this new framework to improve interoperability. FHIR provides definitions for common healthcare concepts that can be accessed and exchanged using the latest web technologies. One common use of this is the patient portal, which is the most frequent way for patients to view their EHRs. With the rise of mobile technology in recent years, people can now access EHRs through mobile apps.

106

Both EHRs and FHIR are real-life examples of data standardization and interoperability in healthcare. They have significantly improved the exchange of healthcare information, leading to improved patient care and reduced costs.

User Feedback and Ethical Considerations

Incorporating user feedback in healthcare IT design is vital for developing systems that meet the real needs of users. Ethical considerations, particularly in the design of user-centric healthcare technology, ensure that the systems are not only efficient but also respectful of patient rights and privacy. Numerous studies have shown that incorporating user feedback in EHR design can lead to a significant reduction in user errors. Ethical considerations, especially in terms of data privacy, are paramount, as concern over health data security is among the highest priorities for the healthcare system.

Digital Inclusivity and Cultural Diversity

Ensuring digital inclusivity and considering cultural diversity are essential for creating healthcare systems that are accessible and useful to all. The importance of designing systems that cater to diverse patient backgrounds and tech-savviness levels cannot be overstated. For example, patient portals designed with inclusivity in mind have a much higher engagement rate among diverse populations.

The future is promising, with artificial intelligence and machine learning expected to play a more significant role. Emerging technologies like virtual reality and augmented reality hold promise for enhancing user experience in healthcare. Future trends likely include further integration of these technologies into standard healthcare practices. Data standardization and interoperability form the backbone of modern healthcare administration. With rigorous standards, innovative practices, and ethical considerations, healthcare systems can balance efficiency, security, and patient-centricity in the digital era.

Future Trends

The landscape of data standardization and interoperability is rapidly evolving, driven by technological advancements and changing regulatory environments. As organizations and industries strive to harness the power of data more effectively, several emerging trends are poised to shape the future of this field.

Blockchain for Data Integrity

Blockchain technology, best known for its role in cryptocurrencies, offers revolutionary potential for ensuring data integrity and security in digital exchanges. By creating decentralized and immutable ledgers, blockchain provides a robust framework for secure and transparent data transactions.

- **Enhanced security**: Blockchain's distributed nature makes data tampering or hacking exceedingly difficult, enhancing the security of data exchanges.

- **Improved transparency:** The technology allows for transparent tracking of data transactions, fostering trust among stakeholders.

- **Decentralization**: By decentralizing data management, blockchain reduces the reliance on central authorities or intermediaries, potentially lowering costs and increasing efficiency in data exchanges.

Artificial Intelligence and Machine Learning

Artificial intelligence (AI) and machine learning (ML) are set to play pivotal roles in enhancing data standardization and interoperability. These technologies can automate complex processes, improve data accuracy, and offer predictive insights.

- **Automated standardization**: AI and ML can automate the cleansing, matching, and standardization of data from diverse sources, significantly improving efficiency and accuracy.

- **Intelligent interoperability solutions**: These technologies can analyze patterns and preferences in data usage and exchange, facilitating the development of more intuitive interoperability protocols and standards.

- **Predictive analytics**: By leveraging historical data, AI can provide predictive insights, enabling organizations to anticipate and adapt to future interoperability challenges and opportunities.

Increased Regulation

As digital data becomes increasingly central to economic and social activities, governments and international bodies are focusing more on data privacy and security. This attention is leading to the formulation of more stringent data protection standards and regulations.

- **Global data privacy laws**: Following the precedent set by the GDPR in Europe, more countries are expected to adopt similar comprehensive data protection laws.

- **Industry-specific regulations**: Sectors handling sensitive information, such as healthcare and finance, may see tighter regulations aimed at enhancing data integrity and protecting consumer information.

- **Harmonization efforts**: There is a growing trend toward the harmonization of data protection laws to simplify compliance for organizations operating internationally. Efforts to standardize regulations can significantly impact data standardization and interoperability practices, making it easier for entities to share data across borders while ensuring privacy and security.

These trends underscore the dynamic nature of data standardization and interoperability. As technologies like blockchain and AI continue to mature and as regulatory landscapes evolve, organizations must remain agile, embracing new tools and adhering to emerging standards to stay competitive in the global digital economy. The future of data management promises enhanced efficiency, security, and innovation driven by these transformative trends.

Conclusion

Data standardization and interoperability are crucial for the seamless exchange and utilization of information across various domains. By embracing these concepts, organizations can enhance their efficiency, foster innovation, and drive forward the digital transformation agenda. As technology evolves, continuous efforts to improve standards and overcome implementation challenges will be key to unlocking the full potential of data in the interconnected world.

PART II

Tools and Experience

CHAPTER 5

Tools and Technologies for Improved Data Usability

In an increasingly data-driven world, the usability of data has become a top priority for organizations seeking to gain a competitive edge. While the sheer volume of data generated by modern systems is unprecedented, the value of this data is not in its quantity but in its usability—how easily it can be accessed, understood, analyzed, and applied to drive informed decision-making.

Usable data is at the heart of the decision-making processes of leading organizations. It is not enough to simply collect vast quantities of data; data must be well-organized, clean, and made accessible to end-users in a format that allows for meaningful analysis. By leveraging the right tools and technologies, companies can improve the usability of their data, ensuring that insights are easily extracted and actions are swiftly taken.

© Saurav Bhattacharya 2025
P. Gujar, *Data Usability in the Enterprise*, https://doi.org/10.1007/979-8-8688-1183-8_5

This chapter explores the essential tools and technologies that enhance data usability, covering topics from data warehousing and integration tools to advanced analytics, machine learning, and collaboration platforms. It also explores future trends, such as augmented analytics and the increasing role of automation in data usability. By understanding and leveraging these technologies, organizations can transform raw data into a strategic asset that drives innovation and operational efficiency.

Defining Data Usability

Data usability refers to the degree to which data can be easily accessed, interpreted, and used for its intended purpose. It ensures that data can be quickly transformed into valuable insights without requiring excessive time, resources, or technical expertise. In essence, usable data has the following characteristics.

- **Accessible**: Data can be retrieved and used by authorized users without significant effort or technical barriers.

- **Understandable**: Data is presented in a format that is easily interpreted by both technical and non-technical stakeholders.

- **Timely**: Data is available when needed, in the right format, and for the right purpose.

- **Reliable**: Data is accurate, consistent, and free from errors, ensuring that decisions based on the data are trustworthy.

- **Interoperable**: Data can be combined, exchanged, and integrated across various systems and platforms.

For data to be usable, it needs to undergo processes that cleanse, organize, and structure it in ways that make it more accessible to a wider range of users. Data usability is essential for organizations that rely on data-driven decision-making because poor data usability can lead to misinformed decisions, inefficiencies, and missed opportunities.

Data Usability in Practice

Imagine a marketing team that wants to analyze customer behavior data to create a targeted campaign. If the data is spread across multiple systems, poorly organized, or full of errors, the team may struggle to extract useful insights. However, if the data is well-integrated, clean, and accessible via a user-friendly interface, the team can easily analyze it to craft a campaign that resonates with the target audience.

The Role of Usability in Data-Driven Systems

Data usability is a key driver of success in modern data-driven systems. As businesses increasingly depend on real-time analytics, big data, and machine learning models, ensuring that data is usable across departments and functions is critical. When data is usable, decision-making becomes faster, more informed, and more strategic.

Usability Enhances Collaboration

In data-driven organizations, data usability is the linchpin of collaboration between various teams. When marketing, sales, operations, and finance teams can access and use the same data seamlessly, the organization operates more efficiently. Cross-functional teams can make better decisions based on shared data, uncovering insights that may have remained hidden within isolated data silos.

Impact on Decision-Making

Usability directly impacts the quality of decisions made. Well-structured, accessible data ensures that decision-makers are acting on complete, accurate information. When data is not usable, decision-making is slowed, and there is an increased risk of errors or missed opportunities due to incomplete or misunderstood data.

Reducing Time Spent on Data Wrangling

Data wrangling—the process of manually cleaning and transforming data into a usable format—is time-consuming and labor-intensive. Improved data usability, through automation and the use of advanced tools, reduces the amount of time spent on these repetitive tasks, freeing up analysts to focus on higher-value work such as uncovering insights and developing strategies.

Tools for Data Collection and Organization

Effective data collection and organization are foundational to creating usable data. Tools that help streamline these processes ensure that data is collected from various sources and made available in a format that is easy to analyze and integrate across systems.

116

Data Warehousing Tools

Data warehousing tools provide centralized repositories for data collected from multiple sources. They offer powerful querying capabilities and support advanced analytics by organizing and storing data in structured formats.

- **Amazon Redshift**, a cloud-based data warehousing solution, allows organizations to run complex queries on large datasets with high performance. Redshift integrates with other Amazon Web Services (AWS) solutions, making it a versatile option for businesses looking to leverage the cloud for data management.

- **Google BigQuery**, part of the Google Cloud Platform, is a serverless, highly scalable data warehouse that allows users to perform fast SQL queries on large datasets. Its integration with machine learning tools makes it a powerful solution for data scientists and analysts.

- **Snowflake**, a multi-cloud data platform, is known for its ability to separate storage and compute resources, providing flexible scaling options. It allows users to quickly access and analyze data from various sources, improving usability for large datasets.

These data warehousing tools enhance usability by offering structured storage for data and enabling fast, efficient access to data through intuitive querying interfaces.

Data Integration and ETL Tools

Data from multiple sources often needs to be integrated into a single platform for analysis. This is where data integration and ETL (extract, transform, load) tools come into play, simplifying the process of combining data from disparate systems and formats.

- **Apache NiFi**, an open-source data integration tool automates the flow of data between systems, offering a highly visual interface for managing large datasets. It enables users to collect, route, and transform data from various sources, improving the usability of large, complex datasets.

- **Talend** provides a suite of tools for data integration and ETL, allowing users to connect and transform data from different platforms, including cloud and on-premises systems. Talend's visual interface makes data workflows more accessible to non-technical users.

- **Informatica PowerCenter**, known for its reliability and scalability, is a robust solution for data integration and ETL. It helps enterprises streamline their data pipelines and manage the flow of data from various sources into their data warehouse.

These tools improve data usability by automating the complex process of extracting, transforming, and loading data into a usable format, freeing users from the need to manually handle disparate data sources.

Tools for Data Visualization and Interpretation

Visualization tools play a critical role in enhancing data usability by converting raw data into visual formats that are easy to understand. When data is presented through charts, graphs, and dashboards, it becomes more accessible to decision-makers, especially those without a technical background.

Visualization Platforms

Data visualization platforms help users create insightful visual representations of data. These platforms improve the interpretability of data and make it easier to detect trends, anomalies, and patterns.

- **Tableau** is a leader in data visualization software, offering users the ability to create interactive visualizations and dashboards without extensive programming knowledge. Its drag-and-drop interface enables users to easily explore datasets and create compelling visual narratives.

- **Microsoft Power BI** is a cloud-based business analytics service that enables users to create and share reports and dashboards. Integrated with the Microsoft ecosystem, it is ideal for businesses that rely on Microsoft products and services.

- **Qlik Sense**, known for its associative data model, allows users to explore data in an intuitive, interactive way. Users can discover insights by exploring associations between different data points, improving their ability to interpret data.

119

Visualization platforms improve data usability by making complex data more accessible through visual storytelling. By presenting data in intuitive formats, these tools empower users to derive insights quickly and confidently.

Dashboard Tools

Dashboards offer real-time insights into key performance metrics and other important data points, making them invaluable tools for decision-making. They consolidate data into a single interface, allowing users to quickly monitor performance, identify trends, and respond to emerging issues.

- **Google Data Studio** is a free data visualization tool that allows users to create customizable, shareable dashboards that integrate with other Google services, such as Google Analytics, Sheets, and Ads. Its simplicity and ease of use make it ideal for small businesses and teams.

- **Klipfolio** provides real-time dashboarding capabilities, enabling organizations to monitor their most important metrics at a glance. The tool integrates with various data sources, including cloud-based services, APIs, and databases.

- **Looker**, part of Google Cloud, is a data exploration and business intelligence platform that offers powerful data visualization and dashboarding capabilities. It allows users to create customized dashboards with real-time data insights.

Dashboards improve data usability by offering a comprehensive, visual view of real-time metrics, enabling decision-makers to monitor and act on key data points quickly.

Data Preparation and Cleaning Tools

Data preparation is a crucial step in making data usable. Raw data often contains errors, inconsistencies, and irrelevant information, making it difficult to analyze. Data preparation tools automate the cleaning, organizing, and transforming of data into a format suitable for analysis, saving time and reducing the likelihood of errors.

Data Wrangling Tools

Data wrangling tools help transform raw data into usable formats by cleaning, restructuring, and integrating it from various sources. These tools enable users to address common data issues such as missing values, inconsistent formats, and duplicated records.

- **Trifacta** uses machine learning to automate many of the steps involved in data preparation, making it easier for users to clean, organize, and transform their data. Its visual interface helps users understand data at a glance and guides them through the preparation process.

- **Dataiku** is an end-to-end data science platform that combines data wrangling, visualization, and machine learning. Its built-in data wrangling tools help users clean and prepare data for analysis, regardless of the source or format.

- **Paxata** provides self-service data preparation tools that allow business users to clean, enrich, and combine data. Its intuitive interface and AI-powered suggestions make it easier for non-technical users to transform raw data into usable datasets.

Data wrangling tools improve data usability by automating the tedious process of cleaning and preparing data, allowing users to focus on analysis and decision-making.

Data Profiling Tools

Data profiling tools help users assess the quality of their data by identifying issues such as missing values, data duplication, and inconsistencies. These tools provide insights into the structure and content of datasets, ensuring that data is clean and reliable before analysis.

- **IBM InfoSphere Information Analyzer** allows organizations to profile, monitor, and understand their data quality. It provides insights into the completeness, accuracy, and consistency of data across various sources.

- **Talend Data Preparation** is a user-friendly platform that enables users to clean and prepare data for analysis quickly. It includes features for detecting and resolving data quality issues, such as missing or inconsistent values.

By providing a clear understanding of data quality, data profiling tools ensure that datasets are accurate and reliable, improving the overall usability of the data.

Machine Learning and AI-Driven Usability Tools

Machine learning and artificial intelligence (AI) technologies are transforming the way organizations interact with and leverage data. AI-driven tools automate data processing, uncover hidden patterns, and provide predictive insights that enhance data usability.

Data Annotation and Labeling

Machine learning models require high-quality, labeled datasets for training. Data annotation and labeling tools are essential for preparing these datasets, ensuring that models are trained on accurate, well-organized data.

- **Labelbox** is a leading platform for data annotation, enabling teams to label, organize, and manage large datasets for machine learning projects. It provides tools for both manual and automated labeling, improving the speed and accuracy of the annotation process.

- **SuperAnnotate** offers AI-powered tools for annotating images and videos, allowing users to label data quickly and efficiently. Its collaborative platform makes it easier for teams to work together on large-scale labeling projects.

Data annotation tools improve the usability of datasets for machine learning models, ensuring that models are trained on high-quality, labeled data.

Predictive Analytics Tools

Predictive analytics tools leverage machine learning algorithms to analyze historical data and make predictions about future outcomes. These tools provide actionable insights that allow businesses to anticipate trends and make proactive decisions.

- **RapidMiner**, an open-source data science platform, allows users to build, train, and deploy machine learning models. Its visual interface makes it accessible to users without extensive coding experience, improving data usability for non-technical teams.

- **H2O.ai** provides a suite of machine learning tools for building and deploying predictive models. Its automated machine learning capabilities automate the machine learning workflow, making it easier for users to build accurate models with minimal effort.

Predictive analytics tools enhance data usability by providing actionable insights based on historical data, enabling organizations to make data-driven decisions with confidence.

AutoML and Smart Assistants for Data Usability

AutoML (automated machine learning) tools simplify the machine learning process by automating the selection of algorithms, hyperparameter tuning, and model training. These tools make machine learning accessible to non-experts, improving the usability of complex data systems.

- **Google AutoML** is a suite of machine learning tools that allow users to build custom models with minimal expertise. It automates many of the tasks involved in building and deploying machine learning models, making data more usable for non-technical teams.

- **DataRobot** provides an automated platform for building and deploying machine learning models. Its AutoML capabilities allow users to quickly develop predictive models without needing in-depth knowledge of machine learning algorithms.

AutoML tools democratize access to machine learning, improving data usability by allowing more users to build and deploy models.

Collaboration and Data Sharing Platforms

Collaboration is key to maximizing the value of data within an organization. Data sharing platforms and collaborative tools ensure that stakeholders can access, share, and analyze data without technical barriers, fostering a more collaborative and data-driven culture.

Cloud-Based Collaboration

Cloud-based platforms enable real-time collaboration on data projects, allowing teams to work together regardless of geographic location. These platforms offer centralized data storage, making it easy for users to access and share data.

- **Google Cloud Platform (GCP)** provides a comprehensive suite of cloud-based tools for data storage, processing, and analysis. Its integration with collaboration tools like Google Drive and Google Docs makes it easy for teams to work together on data projects.

- **Microsoft Azure** offers a range of cloud services for data storage, sharing, and analytics. Its integration with Microsoft Teams and other collaboration tools makes it ideal for organizations using the Microsoft ecosystem.

- **AWS Glue** is a serverless data integration service that makes it easy to move data between different sources and applications. It allows teams to collaborate on data preparation and integration tasks in real time.

These cloud-based platforms improve data usability by enabling teams to collaborate on data projects in real time, regardless of location.

Data Catalogs and Lineage Tools

Data catalogs and lineage tools enhance the discoverability and traceability of data across an organization. These tools provide a centralized repository of metadata that allows users to find and understand relevant datasets.

- **Alation** is a data catalog platform that uses machine learning to improve data discovery and governance. It helps users find, understand, and trust the data they are working with, improving data usability.

- **Collibra** provides data governance and cataloging tools that help organizations manage their data assets. Its data lineage features allow users to trace the origins and transformations of their data, ensuring transparency and trust in data usage.

By improving the discoverability and traceability of data, these tools enhance data usability by making it easier for users to find and trust the data they need.

Data Marketplace Platforms

Data marketplace platforms provide a way for organizations to share and monetize their data with external parties. These platforms improve data usability by allowing businesses to access and integrate third-party datasets, enriching their own data for analysis.

- **AWS Data Exchange** enables organizations to find, subscribe to, and use third-party data in the cloud. It provides access to a wide range of datasets, including financial, healthcare, and retail data.

- **Snowflake Data Marketplace** allows organizations to share and access live, ready-to-query data from third-party sources. It enhances data usability by providing users with a wide range of datasets to enrich their analyses.

Data marketplace platforms improve data usability by expanding the range of data available for analysis, allowing organizations to access valuable third-party datasets.

Technologies Enhancing Data Usability

Emerging technologies like natural language processing (NLP), augmented analytics, and conversational data exploration are transforming how users interact with and derive insights from data. These technologies make data more accessible to non-technical users, enhancing data usability across organizations.

Natural Language Processing

NLP technology allows users to interact with data systems using natural language queries. This eliminates the need for complex query languages like SQL, making data more accessible to non-technical users.

- **Microsoft Power BI Q&A** allows users to ask questions about their data in plain English and receive instant answers in the form of visualizations. This feature makes data analysis more intuitive and accessible to users without technical expertise.

- **Tableau Ask Data** is a feature that allows users to query their data using natural language, providing instant insights without the need for complex SQL queries.

By enabling users to interact with data using everyday language, NLP enhances data usability for a broader audience, allowing more users to access and analyze data.

AI-Powered Querying

AI-powered querying tools use machine learning to automate complex data queries and provide users with recommendations and insights. These tools guide users through the analysis process, making data exploration more intuitive.

- **ThoughtSpot** is an AI-driven analytics platform that allows users to query their data using natural language. It provides instant answers and insights, empowering users to explore their data without needing advanced technical skills.

- **Dremio** is an open-source data lake engine that optimizes data queries using machine learning. It provides users with recommendations for faster query performance, improving the usability of large datasets.

AI-powered querying tools improve data usability by simplifying the querying process and making data exploration more intuitive for non-technical users.

Augmented Analytics

Augmented analytics leverages AI and machine learning to automatically surface insights from data. These tools help users discover patterns and trends they may not have considered, improving the usability of complex datasets.

- **SAS Viya** is a cloud-native analytics platform that uses augmented analytics to provide automatic insights from complex datasets. Its AI-driven tools help users identify trends, outliers, and correlations in their data.

- **Qlik Augmented Intelligence** are augmented analytics tools combine machine learning and AI to assist users in identifying key insights from their data. These tools automatically surface trends and patterns, making data analysis more accessible to a wider audience.

By automating parts of the analysis process, augmented analytics tools improve data usability by lowering the barrier to entry for data exploration and analysis.

Conversational Data Exploration

Conversational data exploration tools use voice or chat interfaces to allow users to interact with data in a conversational manner. These tools make data more accessible by allowing users to ask questions and receive answers through natural language dialogues.

- **Einstein Analytics** by Salesforce offers conversational data exploration capabilities, allowing users to interact with their data using voice commands. This feature enhances the usability of Salesforce data by making it easier for users to retrieve insights through natural language.

- **Siri** offers shortcuts for data exploration. Some analytics platforms integrate with voice assistants like Siri, allowing users to access data and receive insights through voice commands.

Conversational data exploration tools enhance data usability by making it easier for users to retrieve and interact with data through natural, conversational interfaces.

Data Governance Tools for Improved Usability

Effective data governance ensures that data is reliable, secure, and compliant with regulations. Governance tools provide the framework needed to manage data assets across an organization, improving data usability by ensuring that users have access to high-quality, trusted data.

Metadata Management

Metadata management tools help organizations organize and maintain detailed information about their data assets. Metadata provides context about the origin, structure, and use of data, making it easier for users to find and understand relevant datasets.

- **Informatica Metadata Manager** helps organizations manage metadata across their data ecosystems. This tool provides insights into data lineage, helping users understand where their data comes from and how it has been transformed over time.

- **Apache Atlas** is an open-source metadata management platform that helps organizations manage their data assets. It provides tools for managing metadata, data lineage, and data governance policies.

By providing users with detailed information about their data, metadata management tools improve data usability by enhancing the discoverability and transparency of data assets.

Privacy and Compliance

As data privacy regulations become more stringent, ensuring compliance with these regulations is critical for data usability. Privacy and compliance tools help organizations manage sensitive data and ensure that it is handled in accordance with regulatory requirements.

- **BigID** is a data privacy and protection platform that helps organizations discover, classify, and manage sensitive data. It provides tools for ensuring compliance with data privacy regulations and protecting sensitive information.

- **OneTrust** offers tools for managing data privacy, governance, and compliance with global regulations. Its platform helps organizations ensure that their data is handled in accordance with privacy laws.

Privacy and compliance tools improve data usability by ensuring that users can trust the data they are working with, knowing that it is compliant with relevant regulations.

Data Quality and Auditing Tools

Data quality tools ensure that data is accurate, complete, and free from errors, improving its usability. These tools help organizations monitor and maintain the quality of their data assets.

- **Talend Data Quality** is a tool that helps organizations assess and improve the quality of their data. It provides features for detecting and resolving data quality issues, such as missing or inconsistent values.

- **Informatica Data Quality** is a comprehensive solution for monitoring and improving data quality across an organization's data ecosystem. It provides tools for profiling, cleansing, and standardizing data.

By ensuring that data is accurate and reliable, data quality tools improve data usability by providing users with high-quality datasets for analysis and decision-making.

Open-Source Tools and Frameworks

The open-source community has developed a wide range of tools that enhance data usability. These tools are highly customizable, allowing organizations to tailor them to their specific needs without incurring high licensing costs.

Data Manipulation Libraries

Open-source data manipulation libraries enable developers and data scientists to efficiently clean, transform, and analyze data. These libraries provide powerful tools for working with structured data in various formats.

- **Pandas** is a popular Python library for data manipulation and analysis. It provides data structures and functions for working with structured data, making it a powerful tool for data cleaning and transformation.

- **Dask** is a parallel computing library that extends Pandas to handle larger datasets across distributed systems. It provides tools for parallelizing data manipulation tasks, improving the scalability of data workflows.

These open-source libraries improve data usability by providing flexible and powerful tools for data manipulation and analysis.

Visualization Libraries

Open-source visualization libraries enable developers to create custom data visualizations and dashboards. These libraries provide powerful tools for creating interactive and static visualizations in various programming languages.

- **Matplotlib** is a widely used Python library for creating static, animated, and interactive plots. It provides a flexible and customizable platform for visualizing data in various formats.

- **D3.js** is a JavaScript library that allows developers to create highly customizable and interactive data visualizations. Its flexibility and power make it a popular choice for creating advanced visualizations for the web.

These open-source libraries improve data usability by providing flexible tools for creating custom visualizations and dashboards.

Distributed Processing Frameworks

Distributed processing frameworks enable organizations to process large datasets across distributed systems. These frameworks provide the scalability needed to handle big data workflows, improving the usability of large datasets.

- **Apache Spark** is a distributed computing framework that provides fast, scalable data processing capabilities. It supports batch and real-time processing, making it a versatile tool for big data workflows.

- **Hadoop** is an open-source framework for processing and storing large datasets across distributed systems. Its ecosystem includes tools for distributed storage, data processing, and data querying.

These distributed processing frameworks improve data usability by providing the scalability needed to process large datasets efficiently.

Future Trends in Data Usability

As data continues to grow in volume and complexity, new trends are emerging that aim to further enhance data usability. Automation, real-time analytics, and AI-driven tools are shaping the future of how users interact with data.

Automation in Data Usability

Automation is playing an increasingly significant role in improving data usability. Tools that automate data cleaning, preparation, and analysis allow users to focus on extracting insights rather than performing manual tasks.

AutoML platforms like Google AutoML and DataRobot are automating the machine learning pipeline, making it easier for non-experts to build and deploy models.

Automation tools improve data usability by reducing the manual effort involved in data preparation and analysis, allowing users to focus on higher-value tasks.

Real-Time Data Usability Tools

Real-time data is becoming more critical in industries such as finance, healthcare, and e-commerce. Real-time analytics platforms provide immediate insights, allowing users to react to events as they happen.

- **Apache Kafka** is a distributed streaming platform that processes real-time data streams, enabling organizations to analyze data as it is generated.

- **Google BigQuery** offers real-time analytics capabilities that allow users to run queries on streaming data, providing immediate insights from live datasets.

Real-time data usability tools improve decision-making by providing users with up-to-the-minute insights from their data.

Federated Learning and Data Usability

Federated learning is an emerging trend in machine learning that allows models to be trained across decentralized data sources. This approach enhances data usability by enabling organizations to leverage distributed datasets for machine learning without requiring centralized data storage.

- **Google** is a pioneer in federated learning, using this approach to train machine learning models across decentralized devices while preserving user privacy.

- **IBM** is also exploring federated learning, developing tools that allow organizations to train machine learning models across multiple data sources without transferring sensitive data.

Federated learning improves data usability by allowing organizations to leverage distributed data sources while maintaining data privacy and security.

Tools and Technologies for Improved Data Usability in Healthcare

Data usability is a critical aspect of not only clinical research and patient-provider interaction but also healthcare administration and management. It refers to the ease of use, accessibility, and comprehensibility of data. In the healthcare sector, data usability can significantly impact patient outcomes, operational efficiency, and overall quality of care.

Digital transformation, on the other hand, is the integration of digital technology into all areas of a business, fundamentally changing how you operate and deliver value to customers. In healthcare, digital transformation can enhance data usability by improving data collection, storage, analysis, and sharing processes.

The intersection of data usability and digital transformation in healthcare is a dynamic and rapidly evolving field. This chapter explores the tools and technologies that are driving this transformation and improving data usability in healthcare administration and management.

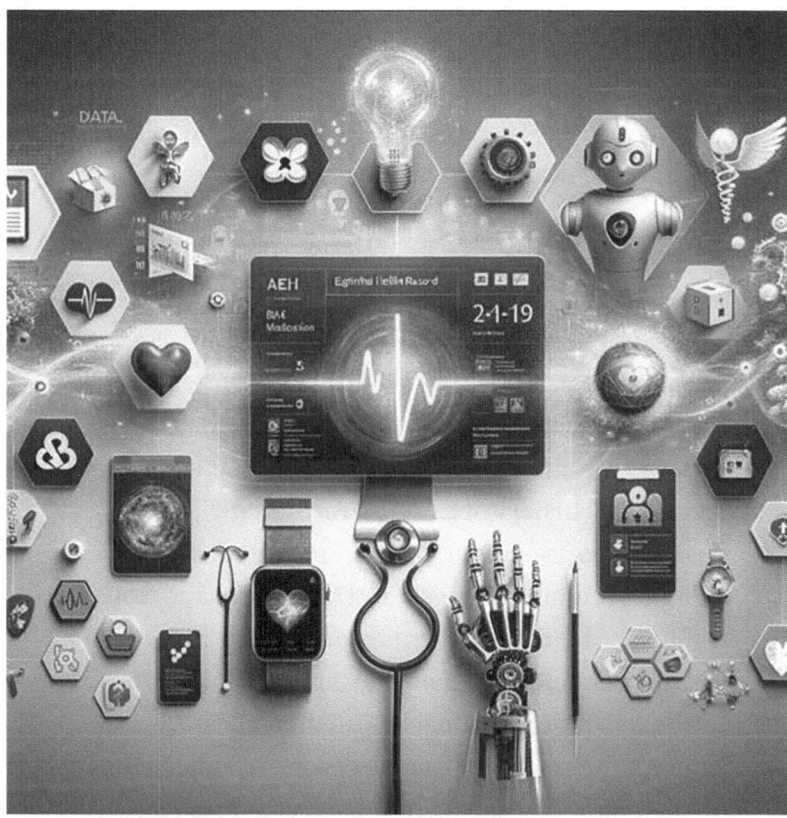

Image Source: DALL·E 2024-03-28 00.29.31

Digital Transformation in Healthcare

Digital transformation in healthcare is a broad and multifaceted concept. It encompasses everything from electronic health records and telemedicine to AI and machine learning.

These technologies are revolutionizing healthcare administration and management by making data more accessible, understandable, and actionable. For instance, electronic health records allow the seamless sharing of patient data across different healthcare providers, improving coordination of care and patient outcomes. Similarly, AI and machine learning can analyze vast amounts of data to uncover patterns and insights that can inform decision-making and improve patient care. The next section delves deeper into some of these tools and technologies and their role in improving data usability in healthcare.

Tools and Technologies in Healthcare

Several tools and technologies can improve data usability in healthcare.

- **Electronic health records (EHRs)** are digital versions of patients' paper charts. They contain all the medical history of patients, including diagnoses, medications, treatment plans, immunization dates, allergies, radiology images, and laboratory and test results. By digitizing patient records, EHRs make data easily accessible and shareable among healthcare providers, improving the coordination of care and allowing for more informed decision-making. EHRs can also support other care-related activities directly or indirectly through various interfaces, including evidence-based decision support, quality management, and outcomes reporting.

- **Data analytics** in healthcare involves the use of statistical techniques to analyze vast amounts of healthcare data and extract meaningful insights. These insights can inform policy-making, improve operational efficiency, and enhance patient care. For instance, predictive analytics uses historical data, machine learning, and statistical algorithms to predict future outcomes. It can help healthcare providers identify at-risk patients and provide preventive care. Prescriptive analytics suggests decision options on how to take advantage of a future opportunity or mitigate a future risk and shows the implication of each decision option. It can support clinical decision-making and personalized medicine.

- **Wearable devices** such as fitness trackers and smartwatches are playing an increasingly important role in healthcare. They allow the continuous monitoring of a patient's health and can provide valuable data for preventive care and disease management. For example, wearable devices can track a patient's heart rate, sleep patterns, activity levels, and more. This data can be analyzed to gain insights into a patient's health and behavior, inform treatment plans, and monitor the progress of treatment. Moreover, as these devices become more advanced, they are being used to manage chronic conditions like diabetes by continuously monitoring glucose levels.

- **AI**, particularly machine learning, is revolutionizing healthcare by providing predictive analytics, precision medicine, and personalized patient care. Machine learning algorithms can analyze vast amounts of healthcare data to predict disease outbreaks, identify at-risk individuals, and even suggest personalized treatment plans. For instance, AI algorithms can analyze patient data to predict the likelihood of readmission, helping healthcare providers to intervene early and prevent unnecessary hospitalizations.

- **Big data** in healthcare refers to the vast quantities of data—created by EHRs, wearable devices, and other sources—that are too large or complex to be processed by traditional data processing software. With the right tools, this data can be analyzed to uncover patterns, trends, and insights that can improve patient care. For instance, by analyzing patient data, healthcare providers can identify trends that might indicate the onset of a disease.

- **3D printing** is being used in healthcare to create patient-specific models for surgical planning, to produce custom prosthetics and implants, and even to print human tissue. For example, surgeons can use 3D-printed models of a patient's anatomy to plan surgeries, which can lead to more precise and effective treatment.

- **Blockchain** technology can enhance data security and patient privacy in healthcare. By creating a decentralized and immutable ledger of health records, blockchain can ensure the integrity and security of patient data while also giving patients more control over their information. For instance, blockchain could be used to create a secure, interoperable health data exchange that enhances collaboration among healthcare providers and improves patient care.

Each of these technologies has the potential to transform and significantly improve data usability in healthcare, leading to better patient outcomes, more efficient operations, and improved quality of care. However, their implementation also requires careful consideration of factors such as data privacy and security, regulatory compliance, and technological infrastructure.

Intelligent Data Management in Healthcare

Let's analyze the usage of one such technology, which helped improve decision-making and led to better data usability.

In their 2022 paper, "Intelligent Data Management in Healthcare" (see `https://ieeexplore.ieee.org/document/9765260`), Pathapati and Gochhait explore the role of intelligent data management in facilitating decision-making in healthcare. They emphasized the importance of efficiently managing structured and processed data, which is crucial in the healthcare sector, where vast amounts of data are generated daily. The authors utilized VOSviewer software for their analysis, a tool commonly used for constructing and visualizing bibliometric networks. These networks can include journals, researchers, or individual publications, and they provide a visual representation of relationships between these entities.

In the context of their study, the authors mapped keywords to various fields impacting data management in healthcare. This approach allowed them to identify key areas of focus and concern in the realm of healthcare data management, providing a comprehensive overview of the field. The resulting framework can aid healthcare professionals and decision-makers in understanding and navigating the complexities of data management in healthcare. By highlighting the most relevant fields and topics, it served as a guide for future research and policy-making.

The authors' work underscored the potential of intelligent data management in improving healthcare outcomes. By leveraging advanced tools and techniques, healthcare providers can make more informed decisions, ultimately leading to better patient care.

Digital transformation is indeed revolutionizing healthcare administration and management by enhancing data usability. Tools and technologies like EHRs, data analytics, wearable devices, AI, big data, 3D printing, and blockchain are at the forefront of this transformation. They are not only making data more accessible and manageable but also enabling more informed decision-making, thereby enhancing patient care and operational efficiency. As these technologies continue to evolve and mature, they are expected to play an increasingly vital role in healthcare, driving further improvements in data usability and ultimately leading to better healthcare outcomes.

Conclusion

The usability of data is a critical factor in the success of any data-driven organization. By leveraging the right tools and technologies—ranging from data collection and cleaning tools to advanced machine learning platforms and real-time analytics—organizations can ensure that their data is accessible, reliable, and actionable.

Improved data usability leads to faster decision-making, better collaboration, and more accurate insights. As technologies such as AI, NLP, and automation continue to evolve, the future of data usability looks even brighter, with tools becoming more intuitive and accessible to a broader range of users.

By investing in tools that enhance data usability, organizations can unlock the full potential of their data, driving innovation, efficiency, and competitive advantage in a rapidly changing business landscape.

CHAPTER 6

Data Visualization: Techniques and Best Practices

In today's data-rich world, making sense of vast quantities of information is a critical challenge. The value of data lies not just in its volume but in the insights that can be drawn from it. Data visualization is the graphical representation of information and data, enabling viewers to understand patterns, trends, and outliers in complex datasets. By transforming raw data into visual form—such as charts, graphs, maps, or infographics—data visualization helps users see the meaning behind the numbers and make informed decisions.

Data visualization is not just about making data look appealing; it is about providing a visual context that allows users to interpret the data intuitively. With effective visualization, complex information is easier to digest, and actionable insights can be quickly identified. For businesses, governments, researchers, and individuals, effective data visualization is a powerful tool to communicate information clearly and efficiently.

© Saurav Bhattacharya 2025
P. Gujar, *Data Usability in the Enterprise*, https://doi.org/10.1007/979-8-8688-1183-8_6

As data continues to grow exponentially in volume and complexity, the importance of data visualization also increases. The tools and techniques used to visualize data have evolved, enabling individuals to create both static and interactive visualizations that can be explored in real time. This chapter discusses various data visualization techniques, tools, and best practices that help transform data into insights, improve communication, and support decision-making.

The Importance of Effective Data Visualization

Data visualization plays a pivotal role in extracting value from data by presenting it in a visually compelling format. The human brain is better equipped to process visual information than raw numbers or text, and a well-crafted visualization can reveal insights that would otherwise be difficult to discern.

The following explains some reasons why data visualization is so critical.

Enhanced Understanding and Insights

Data visualization simplifies complex datasets by organizing information in a clear and digestible manner. This allows individuals to quickly grasp relationships, trends, and outliers. For example, a line chart showing sales over time can immediately highlight peaks and troughs, helping decision-makers understand the data at a glance. Visualizations make it easier to see the "big picture" while also allowing users to drill down into specific data points for deeper analysis.

Improved Decision-Making

In the business world, data-driven decision-making is paramount. Whether it's tracking sales performance, monitoring customer satisfaction, or analyzing market trends, data visualizations enable stakeholders to make informed decisions quickly. A clear visualization reduces ambiguity, highlights critical insights, and provides a solid foundation for action. The ability to interpret data efficiently and make quick decisions is often a competitive advantage for businesses.

Engagement and Communication

Data visualizations are inherently more engaging than large tables of numbers or text-heavy reports. People are drawn to visuals, and a well-designed visualization can capture and hold attention more effectively than raw data. Visuals also break down language barriers, making it easier to communicate complex information to a diverse audience. For example, an infographic summarizing a company's financial performance might be more compelling and easier to understand than a lengthy financial report.

Storytelling with Data

Effective data visualization tells a story, guiding the viewer through the data and helping them draw conclusions. A well-designed chart or graph is more than just a representation of numbers; it can communicate a narrative that leads to greater understanding. For instance, a time-series chart showing the impact of marketing efforts on customer acquisition over time can tell the story of how different campaigns influenced growth. This makes the data relatable and actionable.

Pattern and Outlier Detection

One of the greatest strengths of data visualization is its ability to reveal patterns, trends, and outliers in data. Through visual representation, analysts can spot anomalies or deviations that might indicate problems, opportunities, or unexpected behaviors. Scatter plots, for example, are excellent for identifying correlations or detecting outliers that could otherwise be missed in tabular data.

Types of Data Visualizations

There is a wide range of data visualization types available, each serving a different purpose. Choosing the correct type of visualization depends on the nature of the data and the message you wish to communicate. This section covers several commonly used types of visualizations and their applications.

Charts and Graphs

Charts and graphs are some of the most widely used visualizations for representing numerical and categorical data. They make it easy to compare values, observe trends, and communicate relationships.

- **Bar charts** are the most often used chart type and are ideal for comparing categorical data across different groups. They can be used horizontally or vertically, with each bar representing a category and the length or height corresponding to a value.

- **Line charts** are perfect for displaying trends over time. They are particularly useful when tracking continuous data, such as stock prices or temperature changes. Line charts help users identify trends, patterns, and fluctuations.

146

- **Pie charts** are effective for showing proportions or percentages within a whole. Each slice of the pie represents a category's contribution to the total. However, they are best used for small datasets with limited categories, as too many slices can make the chart hard to read.

- **Scatter plots** display the relationship between two numerical variables by plotting individual data points. They are commonly used to identify correlations, trends, or outliers in datasets. Scatter plots can also be enhanced with color and size to show additional dimensions.

- **Histograms** are similar to bar charts but are used to represent the distribution of continuous numerical data. They group data into bins and show how frequently data points fall into each range. Histograms are particularly useful for identifying the shape of the data distribution.

- **Bubble charts** are an extension of scatter plots, where the size of the bubble represents an additional dimension. These charts are often used to show relationships between three variables, with position determined by two variables and size representing a third.

Maps

Maps are essential for visualizing data that is tied to geographical locations. Geographical data visualizations help users see patterns and trends across regions, and they are commonly used in applications such as population studies, sales analysis, and environmental monitoring.

- **Choropleth maps** use color gradients to represent data values across different geographical areas, such as countries, states, or districts. For example, a choropleth map can be used to show population density or income levels by region.

- **Heat maps** display data density or intensity in a specific geographic area using color shading. They are often used to show the concentration of events or occurrences, such as crime rates or website visitor activity.

- **Dot distribution maps** use dots to represent the frequency or quantity of a variable in a specific location. Each dot represents a certain number of occurrences, allowing viewers to understand geographic distribution.

- **Flow maps** visualize the movement of objects or entities from one location to another, such as migration patterns, shipping routes, or traffic flows. They are often used to depict how people or resources move geographically over time.

Tables

Although not as visually engaging as charts or graphs, tables are an essential part of data visualization. Tables organize data into rows and columns, making them ideal for displaying detailed data and allowing for easy comparison across multiple variables. Tables are most effective when numerical accuracy is important, such as financial reports or scientific data analysis.

Infographics

Infographics combine text, images, and data visualizations to present information in a compelling and easy-to-digest format. They are commonly used to summarize complex data in a visual format that can be shared with a wide audience. Infographics can integrate various visual elements, such as charts, icons, and illustrations, to tell a data-driven story. They are often used in marketing, journalism, and education.

Principles of Effective Data Visualization

Effective data visualization is about more than just creating attractive visuals. It requires thoughtful consideration of how the data is represented, ensuring that it communicates the right message clearly and accurately.

The following describes the core principles that should guide any data visualization project.

Clarity

The primary goal of data visualization is to make data easier to understand. Clarity ensures that the visualization is free from unnecessary complexity and that users can quickly interpret the data being presented. Avoid cluttering visualizations with excessive elements, such as too many labels, gridlines, or decorative elements. The simpler the design, the more effective the visualization.

Accuracy

A data visualization must faithfully represent the data without distortion. Accuracy is critical because misleading visualizations can lead to incorrect conclusions. For example, manipulating the scale of a graph to exaggerate differences between data points can mislead viewers. It is important to maintain the integrity of the data and ensure that the visualization accurately reflects the underlying information.

Relevance

The type of visualization chosen should match the nature of the data and the insights you are trying to convey. For instance, if you are showing trends over time, a line chart is more appropriate than a pie chart. Choosing the right type of visualization ensures that the data is communicated effectively and that the viewer can extract relevant insights.

Consistency

Consistency in design helps viewers make sense of the data. Use consistent color schemes, fonts, and chart types throughout your visualizations. For example, if you are comparing multiple datasets, use the same color scheme for the same categories across all charts. Consistency also applies to axis scales and formatting, ensuring that users do not need to recalibrate their understanding of the data for each visualization.

Aesthetics

While the primary purpose of data visualization is to communicate information, aesthetics play an important role in user engagement. A visually appealing design captures attention and makes the data more memorable. However, aesthetics should never come at the expense of clarity or accuracy. Visual embellishments should be subtle and used to enhance, not distract from, the data.

Best Practices for Choosing the Right Visualization

Selecting the appropriate type of visualization is one of the most important decisions when designing data visualizations. The right choice can make complex data more understandable, while the wrong choice can obscure important insights.

The following explains best practices for choosing the right visualization.

Understand the Nature of Your Data

The structure of your data—whether it's categorical, numerical, temporal, or geographical—should guide your choice of visualization. For example, line charts are ideal for temporal data (data over time), while bar charts are more suited for comparing categorical data.

- **Temporal data**: Best visualized with line charts, area charts, or Gantt charts to show changes over time.

- **Categorical data**: Bar charts or stacked bar charts are ideal for comparing different categories or groups.

- **Geographical data**: Maps are the go-to visualization for spatial data, with choropleth maps, dot maps, and heat maps being common choices.

- **Hierarchical data**: Treemaps, sunburst charts, and dendrograms are effective for showing relationships within hierarchical data structures.

Consider the Purpose of the Visualization

The goal of your visualization influences your choice. Are you trying to compare values, show a relationship between variables, display a distribution, or visualize a trend?

- **Comparison**: Bar charts, grouped bar charts, and column charts are effective for comparing values across categories.

- **Distribution**: Histograms, box plots, and violin plots are ideal for showing the distribution of data.

- **Correlation/relationship**: Scatter plots, bubble charts, and pair plots are useful for visualizing relationships between variables.

- **Composition**: Pie charts, waterfall charts, and stacked bar charts can show how different components contribute to a whole.

Simplify When Possible

Sometimes, the simplest visualizations are the most effective. If a bar chart communicates the insight clearly, there is no need to use a more complex visualization, like a 3D pie chart. Avoid adding complexity for the sake of aesthetics.

Tailor to Your Audience

Consider the level of expertise and the needs of your audience. A technical audience might appreciate a more complex and detailed visualization, such as a scatter matrix or a heat map. In contrast, a non-technical audience may prefer simpler charts like bar or line charts that communicate key insights quickly.

Use Multiple Visualizations for Complex Data

For datasets with multiple variables or dimensions, it may be necessary to use multiple visualizations to communicate the full picture. For example, a combination of a line chart and a bar chart may be used to show trends over time, along with category comparisons. Dashboards are an excellent way to present multiple related visualizations in one cohesive view.

Designing for Clarity and Readability

The effectiveness of a data visualization hinges on how easily the viewer can interpret the information. Good design practices enhance clarity and readability, allowing users to extract insights quickly.

The following discusses key design elements to consider.

Use of Color

Color is a powerful tool in data visualization, but it must be used strategically. While color can highlight key insights and differentiate between data points, misuse of color can lead to confusion or misinterpretation.

- **Limit the color palette.** Too many colors can overwhelm the viewer. Stick to a limited color palette that highlights key data points without causing visual noise.

- **Use color scales for quantitative data:** Use color gradients to represent numerical ranges. For example, choropleth maps often use gradients from light to dark to show varying intensities of a data variable (e.g., population density).

- **Use contrast for emphasis.** High contrast between elements, such as using a bright color against a dark background, can emphasize critical data points or trends. However, avoid excessive contrast that can strain the eyes.

- **Avoid relying solely on color.** Since some viewers may have color blindness, do not rely solely on color to differentiate data. Incorporate patterns, labels, or icons to provide additional visual cues.

- **Cultural considerations**: Colors carry different meanings in different cultures. For instance, red may symbolize danger or a negative outcome in some cultures, while it may represent prosperity in others. Be mindful of the cultural context when designing visualizations.

Labels and Annotations

Labels and annotations play a crucial role in providing clarity to data visualizations. Without proper labels, even a well-designed visualization can become confusing or ambiguous.

- **Axis labels**: Always label the axes in graphs and charts to ensure viewers understand the context of the data. Without clear axis labels, users might misinterpret the relationship between variables.

- **Direct labeling**: Whenever possible, label data points directly rather than using a separate legend. This reduces the cognitive load of having to cross-reference between the chart and the legend.

- **Annotations for key insights**: Use annotations to highlight significant data points or trends. Annotations help guide the viewer's attention to key findings and can provide additional context or explanations.

- **Legends and keys**: If your visualization requires a legend (e.g., for color coding), place it in a location where it is easy to find but doesn't obscure the data. Ensure that the legend is clear and concise.

Typography

Typography plays an important role in data visualization, affecting both readability and aesthetics. Poor typography choices can make even the most well-designed visualization difficult to understand.

- **Font size**: Ensure that text is large enough to be readable on different devices but not so large that it overwhelms the visualization. The size of labels, axis titles, and annotations should be proportionate to the visualization.

- **Font style**: Stick to simple, sans-serif fonts for clarity. Avoid overly decorative fonts, which can be distracting or hard to read. Consistency in font style throughout the visualization also enhances readability.

- **Use of hierarchy**: Create a hierarchy in typography to guide the viewer's attention. For example, use a larger, bolder font for main titles and a smaller font for axis labels or captions. This helps viewers focus on the most important information first.

- **Legibility across devices**: With visualizations often being viewed on a range of devices—from desktops to smartphones—ensure that your typography remains legible regardless of screen size.

Interactivity in Data Visualization

Interactive data visualizations empower users to explore data on their own terms, enabling deeper engagement and a more personalized experience. Unlike static visualizations, interactive visualizations allow users to manipulate the data, revealing insights that may not be immediately apparent in a static chart.

Drill-Down and Exploration

Interactive visualizations allow users to drill down into the data by clicking on specific elements, such as bars, lines, or regions on a map, to reveal more detailed information. For example, a sales dashboard might allow users to click on a particular product category to see detailed sales data for each product within that category. This ability to explore data at multiple levels provides a richer understanding of the data.

Filters and Customization

Filters enable users to adjust the data displayed in the visualization based on specific criteria. Common filters include date ranges, categories, and geographic regions. By allowing users to customize the data they view, filters provide flexibility and ensure that the visualization remains relevant to the user's needs.

For example, a marketing dashboard might allow users to filter website traffic data by geographic location, device type, or referral source. This helps users focus on the data that is most important to them.

Hover Information and Tooltips

Tooltips provide additional information when a user hovers over a specific data point, allowing for a more detailed view of the data without cluttering the main visualization. For example, in a scatter plot, hovering over a point might display details such as the exact x and y values or a brief description of the data represented by the point.

Tooltips are particularly useful for providing context without overwhelming the user with too much information upfront. They make it easy for users to explore data interactively and extract deeper insights.

Responsive Design

Interactive visualizations should be designed to work seamlessly across different devices and screen sizes. Responsive design ensures that visualizations adjust to fit various screen resolutions, from desktop monitors to mobile phones. This is especially important as more users access dashboards and reports on mobile devices.

Interactive Dashboards

Interactive dashboards bring multiple visualizations together in one cohesive interface, allowing users to interact with and analyze data from different angles. Dashboards often include filters, drill-down capabilities, and dynamic updates, making them ideal for real-time data analysis and decision-making.

For example, a financial dashboard might include line charts showing stock prices over time, bar charts comparing company performance, and tables displaying key financial metrics. Users can interact with each of these elements to customize their view and focus on the data that matters most to them.

Interactive dashboards are particularly valuable for business intelligence and performance tracking, as they provide a comprehensive view of key metrics in real time.

Data Visualization Tools and Software

The growing demand for data visualization has led to the development of a wide range of tools and software solutions designed to help users create high-quality visualizations quickly and efficiently. These tools range from simple charting tools to powerful platforms for creating interactive dashboards.

The following is an overview of some of the most commonly used data visualization tools.

Tableau

Tableau is a very popular tool for creating interactive and shareable data visualizations. It allows users to build sophisticated dashboards that can be shared with others or embedded in web applications. Tableau is known for its user-friendly interface, making it accessible to both technical and non-technical users. Its drag-and-drop functionality allows users to create visualizations without needing to write code.

Tableau integrates with a wide range of data sources, including databases, spreadsheets, and cloud-based services. It offers advanced features for data exploration, including filters, drill-down capabilities, and real-time updates, making it a powerful tool for business intelligence and analytics.

Power BI

Microsoft Power BI is a business analytics tool that enables users to create interactive reports and dashboards. Like Tableau, Power BI is designed to be user-friendly, with drag-and-drop functionality that allows users to create visualizations without coding.

Power BI integrates seamlessly with Microsoft products such as Excel, Azure, and SharePoint, making it an attractive option for organizations that already use Microsoft tools. It also supports a wide range of data sources, including cloud-based services, databases, and Excel spreadsheets.

Power BI offers robust sharing and collaboration features, allowing teams to work together on reports and dashboards in real time. It also includes AI-powered insights that can help users identify patterns and trends in their data.

D3.js

D3.js (data-driven documents) is a powerful JavaScript library for creating custom data visualizations in web browsers. Unlike tools like Tableau or Power BI, which offer a graphical user interface, D3.js requires coding knowledge and is typically used by developers and data scientists who want to create highly customized visualizations.

D3.js allows users to manipulate data and generate dynamic, interactive visualizations that are rendered directly in the web browser. It is highly flexible, supporting a wide range of chart types and interaction models. While D3.js has a steep learning curve, its flexibility makes it a popular choice for creating bespoke visualizations.

Google Data Studio

Google Data Studio is a free tool that allows users to create interactive reports and dashboards. It integrates with a wide range of Google services, including Google Analytics, Google Ads, and Google Sheets, making it a popular choice for marketers and analysts who work with Google products.

Google Data Studio is designed for ease of use, with a drag-and-drop interface that allows users to create visualizations without needing to write code. It also supports real-time data updates, making it a valuable tool for monitoring and analyzing data as it changes over time.

R and ggplot2

For users with a background in programming and statistical analysis, the R programming language offers powerful tools for data visualization. R's ggplot2 package is one of the more popular libraries for creating high-quality statistical graphics. ggplot2 allows users to create a wide range of charts and graphs, from simple bar charts to complex multidimensional visualizations.

The power of ggplot2 lies in its flexibility and customization options. Users can control every aspect of the visualization, from the type of chart to the colors, labels, and annotations. This makes it an ideal tool for data scientists and statisticians who need to create detailed, precise visualizations for research and analysis.

Common Mistakes in Data Visualization

Even experienced designers can make mistakes when creating data visualizations. These mistakes can lead to confusion, misinterpretation, or even mistrust in the data being presented.

The following covers some of the most common mistakes to avoid in data visualization.

Misleading Scales

A common mistake in data visualization is manipulating the scale of a graph to exaggerate or minimize differences between data points. For example, if the y-axis of a bar chart does not start at zero, the differences between the bars can appear larger than they actually are, misleading the viewer.

Always ensure that your scales are appropriate for the data being represented. If you need to adjust the scale, clearly explain the reason for doing so in the visualization or accompanying text.

Overcomplicating the Visualization

Too many elements in a visualization—such as excessive data points, colors, or gridlines—can make it overwhelming and difficult to interpret. Simplicity is often key to effective communication. Use only the elements that are necessary to convey the insight, and avoid adding unnecessary complexity for aesthetic reasons.

For example, a pie chart with 15 slices is likely to be confusing and hard to read. A bar chart might be a better option for comparing many categories.

Misusing 3D Visuals

While 3D charts and graphs can be visually striking, they are often harder to read than their 2D counterparts. The use of perspective in 3D charts can distort the data, making it difficult for viewers to accurately compare values.

In most cases, it's best to stick with 2D charts, which are easier to interpret. If you do choose to use a 3D chart, ensure that it adds value to the visualization rather than complicating it.

Ignoring Context

Data visualizations should always include sufficient context to ensure that viewers can interpret the data correctly. This includes clear labels, annotations, and explanations of key findings. Without context, viewers may draw incorrect conclusions from the data.

For example, a line chart showing an increase in sales might appear impressive at first glance, but without knowing the time period, the viewer might not realize that the increase occurred over several years rather than a few months.

Inappropriate Chart Types

Choosing the wrong type of chart can confuse or mislead viewers. For example, using a pie chart to represent changes over time is inappropriate because pie charts are designed to show proportions, not trends. Similarly, using a line chart to compare categorical data can make it harder for viewers to compare values.

Always choose the chart type that best matches the data and the insights you want to communicate.

Advanced Techniques

As data grows in complexity, advanced techniques in data visualization become necessary to effectively communicate insights. This section explores some of the more advanced methods for visualizing complex datasets.

Data Aggregation and Summarization

Large datasets can contain millions or even billions of data points, making it difficult to visualize every individual point. Data aggregation and summarization techniques help reduce the complexity of large datasets by grouping data into meaningful categories or calculating summary statistics.

- **Clustering**: Grouping data points that are similar to each other into clusters can help simplify complex datasets and reveal patterns. Clustering is often used in market segmentation, where customers with similar behaviors are grouped together.

- **Binning**: Binning groups continuous data into intervals, or "bins," to create a simplified view of the data distribution. For example, a histogram uses binning to show the distribution of values across a range.

- **Moving averages**: Moving averages smooth out short-term fluctuations in data, making long-term trends easier to see. They are commonly used in time-series data, such as stock prices or weather patterns.

By aggregating and summarizing data, you can create visualizations that highlight key patterns without overwhelming the viewer with too much detail.

Real-Time Data Visualization

In many industries, real-time data visualization is essential for monitoring performance, tracking trends, and making quick decisions. Real-time visualizations update dynamically as new data is received, allowing users to see the most up-to-date information.

Real-time visualizations are commonly used in fields like the following.

- **Financial markets**: Stock market dashboards display real-time prices, trading volumes, and other key metrics.

- **IoT devices**: Sensor data from Internet of Things (IoT) devices, such as temperature readings or equipment performance, can be visualized in real time to monitor system health and prevent downtime.

- **Web analytics**: Website traffic data, including page views, user sessions, and conversions, can be tracked in real time to optimize marketing efforts and improve user experience.

When designing real-time visualizations, it is important to ensure that the system can handle the data volume and update frequency without compromising performance.

Multivariate Visualizations

Multivariate visualizations display relationships between more than two variables, enabling users to analyze complex datasets and uncover hidden patterns. These techniques allow for the visualization of multiple dimensions within a single chart or graphic.

- **Parallel coordinate plots**: In a parallel coordinate plot, each variable is represented by a vertical axis, and data points are plotted as lines that pass through each axis. This allows users to see how different variables are related across multiple dimensions.

- **Heat maps with multiple dimensions**: Heat maps can be extended to represent more than two dimensions by using color to represent one variable and position to represent another. For example, a heat map can display temperature variations across geographical regions and time periods simultaneously.

- **Bubble charts**: Bubble charts extend the capabilities of scatter plots by using the size of the bubble to represent a third variable. This allows users to analyze the relationship between three variables in one chart.

Multivariate visualizations are powerful tools for data exploration, enabling analysts to uncover correlations, dependencies, and interactions between variables that might not be apparent in simpler visualizations.

Evaluating the Effectiveness of Visualizations

Creating a visualization is only the first step. To ensure that it effectively communicates the intended message, it is important to evaluate its effectiveness.

The following discusses some key criteria for evaluating data visualizations.

Clarity

The most important criterion for evaluating a data visualization is clarity. Can the viewer understand the data being presented without confusion? Does the visualization make the key insights easy to grasp?

A good test for clarity is to show the visualization to someone unfamiliar with the data. If they can quickly understand the main takeaways, the visualization is clear. If they need to ask questions or seem confused, adjustments may be needed.

Accuracy

A visualization must accurately represent the data. Misleading visualizations, whether intentional or not, can result in incorrect conclusions and poor decision-making. Double-check that the data has been plotted correctly, that scales are appropriate, and that any transformations or aggregations are clearly explained.

It's also important to ensure that the visualization does not unintentionally introduce bias. For example, using a pie chart to represent data with small differences between categories can exaggerate those differences, leading to misinterpretation.

Engagement

Engagement refers to how well the visualization captures and holds the viewer's attention. Does the design draw viewers in and encourage them to explore the data further? Visualizations that are aesthetically pleasing, interactive, or that tell a compelling story are more likely to engage users.

Engagement is especially important for dashboards and reports, where users may need to interact with the data regularly. Incorporating interactivity, such as filters or hover information, can enhance engagement and encourage users to explore the data.

Actionability

Ultimately, the goal of data visualization is to enable action. Does the visualization provide insights that are relevant and actionable? For business users, this might mean identifying areas for improvement, spotting opportunities, or making strategic decisions based on the data.

An effective visualization clearly communicates what action needs to be taken and provides the necessary data to support that decision. If the visualization leaves viewers uncertain about what steps to take, it may need to be refined.

Testing and Feedback

One of the best ways to evaluate the effectiveness of visualization is through user testing and feedback. Share the visualization with your target audience and ask for their input. Did they find it easy to understand? Were they able to extract the key insights? Did they have any suggestions for improvement?

Gathering feedback from real users can help you identify areas for improvement and ensure that the visualization meets their needs.

Data Ethics and Visualization

As the use of data visualization becomes more widespread, it is important to consider the ethical implications of how data is presented. Data visualizations have the power to influence decisions, shape opinions, and drive actions, which makes it critical to ensure that they are created and used ethically.

Accuracy and Honesty

The most fundamental ethical responsibility in data visualization is to present the data accurately and honestly. Misleading visualizations can result in poor decisions, loss of trust, and even harm to individuals or organizations. The following are some common ethical issues in data visualization.

- **Manipulating scales**: Altering the scale of a chart to exaggerate or minimize differences can mislead viewers. Always use appropriate scales that accurately represent the data.

- **Omitting key data**: Selectively presenting data that supports a specific narrative while omitting data that contradicts it is unethical. Visualizations should provide a complete and balanced view of the data, allowing viewers to form their own conclusions.

- **Cherry-picking data**: Choosing only the data points that support a particular argument while ignoring the rest of the dataset can distort the true picture. It's important to include all relevant data, even if it doesn't align with the desired narrative.

Privacy and Confidentiality

When working with personal or sensitive data, it is critical to ensure that each individual's privacy and confidentiality are protected. Visualizations should be designed in a way that prevents the identification of individuals or the disclosure of sensitive information.

For example, when visualizing healthcare data, it is important to anonymize patient information and avoid displaying data that could lead to the identification of specific individuals. Similarly, when visualizing customer data, such as purchasing behavior or location data, steps should be taken to protect user privacy.

Transparency

Transparency is essential for building trust in data visualizations. This means being open about where the data comes from, how it was collected, and any transformations or calculations that were applied. Visualizations should also make it clear when data is incomplete or when assumptions have been made.

For example, if a dataset includes estimates or projections, this should be clearly communicated in the visualization or accompanying text. Providing links to the source data or documentation can also help build trust and allow users to verify the accuracy of the data.

Ethical Decision-Making

As data professionals, we have a responsibility to ensure that the visualizations we create are used ethically and responsibly. This includes considering the potential impact of the visualization on decision-making and ensuring that it is used for positive purposes.

For example, a data visualization that highlights economic disparities in different regions can be a powerful tool for driving policy change. However, it's important to consider how the data might be interpreted and whether it could be misused to justify harmful actions.

Accessibility in Data Visualization

Ensuring that data visualizations are accessible to all users, including those with disabilities, is an important consideration in the design process. Accessible visualizations allow everyone, regardless of ability, to understand and interact with the data.

The following discusses best practices for creating accessible data visualizations.

Color Contrast

A common accessibility issue in data visualization is insufficient color contrast. People with visual impairments, including color blindness, may struggle to differentiate between colors that are too similar. To ensure that your visualizations are accessible, use color palettes with high contrast and test them with colorblind-friendly tools.

For example, using a blue-to-orange color gradient instead of red-to-green can help make visualizations more accessible to people with red-green color blindness.

Alt Text for Visuals

Visualizations should include alternative text (alt text) that describes the data being presented. Alt text is especially important for users who rely on screen readers, as it allows them to understand the content of the visualization even if they cannot see it.

When writing alt text, be sure to provide a concise yet detailed description of the key insights from the visualization. For example, instead of simply stating "Bar chart", the alt text might say, "Bar chart showing a 20% increase in sales from 2019 to 2020."

Keyboard Navigation

For interactive visualizations, it's important to ensure that all interactive elements, such as buttons, filters, and hover information, can be accessed using a keyboard. Many users with disabilities rely on keyboard navigation rather than a mouse, so making your visualizations keyboard-accessible ensures that they can interact with the data.

Text Descriptions of Data

In addition to alt text, consider providing a text-based summary of the key findings from the visualization. This summary can be placed below the visualization and should highlight the most important insights, trends, or patterns in the data. Text summaries provide an alternative way for users to access the information and can improve the overall accessibility of the visualization.

Accessible Design Tools

Many modern data visualization tools include features for creating accessible visualizations. For example, Tableau and Power BI offer accessibility features such as screen reader support and colorblind-friendly palettes. Be sure to explore the accessibility features of your chosen tool and incorporate them into your design process.

By making data visualizations accessible, you ensure that they can be understood and used by a broader audience, including individuals with disabilities.

Data Storytelling

Data storytelling is the process of combining data, visuals, and narrative to convey insights in a compelling and impactful way. While data visualization helps communicate patterns and trends in data, data storytelling takes it a step further by crafting a cohesive narrative that guides the viewer through the data and highlights key insights.

The Elements of Data Storytelling

Data storytelling consists of three core elements: data, visuals, and narrative. Each of these elements plays an important role in crafting an effective data story.

- **Data**: The foundation of any data story is the data itself. The data must be accurate, relevant, and reliable. Without strong data, the story lacks credibility.

- **Visuals**: Visualizations bring the data to life and make it easier for viewers to understand. The choice of visualization should align with the data and the insights you want to convey.

- **Narrative**: The narrative ties the data and visuals together into a cohesive story. It provides context, explains the significance of the findings, and helps the viewer understand why the insights matter. The narrative should guide the viewer through the data step by step, building to a clear conclusion.

Crafting a Data-Driven Story

The following steps explain how to craft an effective data story.

1. **Identify the key message.** What is the main insight or conclusion you want the viewer to take away from the data? Start by identifying the key message and build the story around it.

2. **Structure the story.** Like any good story, a data story should have a clear beginning, middle, and end. Start by introducing the data and providing context. In the middle, present the key insights and trends. Finally, conclude by summarizing the findings and explaining their implications.

3. **Choose the right visuals.** The visuals should support and enhance the narrative. Use the principles and best practices outlined in this chapter to choose the most appropriate visualizations for your data.

4. **Simplify the data.** Focus on the most important data points that support your story. Too much data can overwhelm the viewer and dilute the message. Simplifying the data ensures that the story remains clear and focused.

5. **Add annotations and explanations.** Annotations, captions, and explanations help provide context and guide the viewer through the data. Use these elements to highlight key insights and explain the significance of trends or patterns.

6. **Engage the audience.** An effective data story engages the audience by making the data relatable and actionable. Use real-world examples, comparisons, or analogies to help viewers connect with the data.

Examples of Data Storytelling

Data storytelling is used in a wide range of fields, from journalism to business intelligence. The following are a few examples of data storytelling in action.

- **Journalism**: News outlets often use data storytelling to explain complex issues in an engaging way. For example, a news article about climate change might use interactive maps, charts, and narratives to show how global temperatures have risen over time and the impact this has on different regions.

- **Business reporting**: In business, data storytelling is used to communicate performance metrics, trends, and forecasts. A sales report might tell the story of how a new marketing campaign led to an increase in customer acquisition, using charts and graphs to support the narrative.

- **Nonprofits and advocacy**: Nonprofits often use data storytelling to raise awareness about social issues. For example, a nonprofit working to address food insecurity might create a data-driven story that highlights the number of people affected by hunger in different regions and the impact of their programs.

Data storytelling combines the power of data with the art of storytelling to create compelling, informative, and memorable insights.

Dashboards and Reporting

Dashboards are interactive tools that bring together multiple data visualizations into one cohesive interface, allowing users to monitor and analyze key metrics in real time. Dashboards are commonly used in business intelligence, performance tracking, and decision-making.

The Role of Dashboards

Dashboards provide a high-level overview of key metrics, allowing users to quickly identify trends, monitor performance, and make data-driven decisions. Unlike traditional reports, which are static and often outdated by the time they are read, dashboards offer real-time insights that can be updated continuously as new data comes in.

Dashboards are typically used for the following purposes.

- **Performance monitoring**: Dashboards allow users to track key performance indicators (KPIs) in real time. For example, a sales dashboard might display metrics such as revenue, conversion rates, and customer acquisition costs, providing a snapshot of overall performance.

- **Decision-making**: Dashboards help decision-makers understand the current state of the business and identify areas for improvement. For example, a marketing dashboard might show the results of recent campaigns, allowing marketers to adjust their strategies based on the data.

- **Collaboration**: Dashboards enable teams to collaborate on data analysis by providing a shared view of key metrics. Team members can interact with the data, explore different filters, and discuss findings in real time.

175

Designing Effective Dashboards

An effective dashboard should be designed with the user's needs in mind. Here are the best practices for designing dashboards.

- **Keep it simple.** Dashboards should focus on the most important metrics and insights. Avoid cluttering the dashboard with too many charts or data points. Instead, prioritize the metrics that are most relevant to the user.

- **Use consistent layouts.** Consistency in layout and design makes it easier for users to navigate the dashboard and interpret the data. Group related metrics together and use consistent colors, fonts, and chart types.

- **Provide interactivity.** Interactive features such as filters, drill-downs, and hover information allow users to explore the data in more detail. For example, a sales dashboard might allow users to filter the data by region, product category, or time period.

- **Ensure real-time updates.** Dashboards should display the most up-to-date data, especially when used for performance monitoring. Ensure that the data refreshes automatically and that users can see when the data was last updated.

- **Optimize for different devices.** Many users access dashboards on mobile devices, so it's important to ensure that the dashboard is responsive and works well on different screen sizes. Test the dashboard on both desktop and mobile to ensure a seamless experience.

Types of Dashboards

There are several types of dashboards, each designed for a specific purpose.

- **Operational dashboards**: These dashboards provide real-time monitoring of business operations, such as sales, inventory levels, or website traffic. They are designed to provide up-to-the-minute data and are used by managers and teams to track day-to-day performance.

- **Analytical dashboards**: Analytical dashboards allow users to explore data in depth, identifying trends, correlations, and outliers. They often include interactive features such as filters and drill-downs, making them ideal for data exploration and decision-making.

- **Strategic dashboards**: Senior leaders use strategic dashboards to track long-term goals and KPIs. These dashboards focus on high-level metrics and provide a snapshot of overall business performance.

Reporting with Dashboards

In addition to real-time monitoring, dashboards can be used to generate reports. Many dashboard tools, such as Tableau and Power BI, allow users to export dashboards as PDF reports, making it easy to share insights with stakeholders.

For example, a financial dashboard might generate a monthly report that highlights revenue, expenses, and profit margins. The report can be shared with the finance team and executives, providing a clear overview of the company's financial health.

177

By combining real-time monitoring with reporting capabilities, dashboards provide a powerful tool for data analysis, decision-making, and communication.

Conclusion

Data visualization is a critical skill in today's data-driven world. By transforming raw data into visual representations, data professionals can reveal patterns, trends, and insights that would otherwise be hidden. Effective data visualization not only simplifies complex information but also engages the audience, fosters better decision-making, and communicates key insights clearly and effectively.

This chapter explored various techniques and best practices for creating data visualizations, including how to choose the right visualization, design for clarity, and avoid common pitfalls. It also discussed advanced techniques such as real-time data visualization and multivariate visualizations, as well as the ethical considerations involved in working with data.

As data continues to grow in volume and complexity, the ability to create clear, accurate, and compelling visualizations becomes increasingly important. By following the principles and best practices outlined in this chapter, data professionals can create visualizations that not only convey information but also inspire action and drive positive outcomes.

The world of data visualization is constantly evolving, with new tools, techniques, and technologies emerging all the time. Staying up-to-date with the latest developments ensures that you remain at the forefront of this exciting field, ready to turn data into insights and insights into action.

CHAPTER 7

User-Centric Software Strategies

In today's digital age, where users interact with software on an unprecedented scale, usability has become a key differentiator between successful and unsuccessful products. The best technology is only as good as its interface with users, and user satisfaction is a critical factor in adoption, loyalty, and overall success.

User-centric software strategies focus on designing software with a clear understanding of user needs, preferences, and limitations. This chapter provides an in-depth exploration of the principles, methodologies, and tools used to create software that is not only functional but also intuitive and enjoyable to use. By employing user-centric design practices, developers can ensure their software provides a seamless experience for the end-users, resulting in higher satisfaction, reduced errors, and greater efficiency.

Defining Usability in Software

Usability is the degree to which specified users can use a system, product, or service to achieve specified goals with effectiveness, efficiency, and satisfaction in a given context of use. Unlike other design considerations, usability focuses on the human aspect of system interaction—how users perceive and interact with a system.

P. Gujar, *Data Usability in the Enterprise*, https://doi.org/10.1007/979-8-8688-1183-8_7

Key Attributes of Usability

- **Effectiveness**: How accurately and completely users can achieve their goals.

- **Efficiency**: How quickly users can complete tasks with minimal effort or frustration.

- **Satisfaction**: The user's level of comfort and positive experience when interacting with the system.

Factors Influencing Usability

- **Learnability**: How easy is it for users to accomplish basic tasks the first time they encounter the software?

- **Memorability**: When users return to the system after a period of not using it, how easily can they reestablish proficiency?

- **Error tolerance**: How well does the system help users recover from mistakes?

- **Accessibility**: Can the software be used by people with a range of abilities and disabilities?

Understanding these attributes ensures that usability becomes a central pillar in software design, prioritizing the user experience alongside technical functionality.

The Importance of User-Centric Design

In traditional software development, the focus was often on system performance, technical features, and business requirements. However, as technology evolved and users became more sophisticated, user-centric design emerged as a philosophy that places the user at the heart of the design process.

The following are some business benefits of user-centric design.

- **Increased customer loyalty**: Satisfied users are more likely to continue using the product and recommend it to others.

- **Improved product adoption**: An intuitive interface lowers the learning curve, making it easier for new users to get started.

- **Cost savings**: By designing a product that meets user needs from the start, development teams can avoid costly redesigns, reduce customer support, and minimize training costs.

- **Competitive advantage**: Companies that prioritize user experience are more likely to stand out in a crowded marketplace.

- **Enhanced brand reputation**: A product that provides a great user experience reinforces a brand's reputation for quality, reliability, and customer focus.

In an era where users have increasingly high expectations for software, failing to implement user-centric design can result in user frustration, high churn rates, and loss of market share.

Principles of User-Centered Design

User-centered design (UCD) is an iterative design process where designers focus on the users and their needs in each phase of the design process. It involves users from the very beginning and throughout the process to ensure that the final product is usable and meets their needs.

The following describes the core principles of UCD.

- **Context of use**: Understand the users, their goals, and the environments in which they use the product.

- **Involve users**: Engage users early and throughout the process, using feedback to guide design decisions.

- **Iterative process**: UCD is a cycle of designing, testing, and refining. User feedback is used to make continuous improvements.

- **Design for all**: Consider users with diverse abilities, skill levels, and goals. UCD should account for accessibility and usability across a wide range of contexts and users.

- **User feedback and validation**: Regularly gather user feedback and validate assumptions through testing and observation to ensure the design aligns with user needs.

By adhering to these principles, designers, and developers ensure that the product not only functions as intended but also delivers a positive and meaningful experience for the user.

Understanding User Needs: Research and Analysis

Understanding users' needs is the cornerstone of user-centric software strategies. Before any design work begins, research must be conducted to gather insights about the users, their goals, behaviors, and challenges.

Methods of User Research

- **User interviews**: In-depth, one-on-one discussions that allow researchers to explore users' experiences, needs, and pain points.

- **Surveys and questionnaires**: These tools help gather quantitative data from a large number of users, identifying trends and common challenges.

- **Contextual inquiry**: This involves observing users in their natural environment to see how they interact with existing systems and where inefficiencies occur.

- **Focus groups**: Groups of users discuss their experiences with a product, providing qualitative data about their needs, frustrations, and expectations.

- **Usability audits**: Review current systems or competitor products to identify gaps in usability and areas for improvement.

Analyzing User Data

The data gathered from research should be synthesized into actionable insights. Patterns in user behavior, recurring pain points, and common goals should inform design decisions. **Affinity diagrams** are a common method of organizing research data into meaningful categories that highlight user priorities and challenges.

Understanding user needs helps to create a design that is purposeful and aligned with user goals, resulting in a more intuitive and user-friendly product.

Personas and User Stories

Personas and user stories are essential tools in user-centric design. They help designers and developers stay focused on user needs and use cases throughout the design and development process.

Personas

A **persona** is a fictional representation of a target user created based on research and real data. It helps to humanize the user and create a shared understanding across the team of who they are designing for. Each persona includes the following.

- **Demographic information**: Includes age, occupation, background, and so forth.

- **Goals and motivations**: What is the user trying to accomplish? What drives them?

- **Frustrations and pain points**: What are the user's challenges or frustrations with current systems?

- **Technical proficiency**: How comfortable is the user with technology? What is their experience level?

- **Use case scenarios**: Descriptions of the tasks the persona needs to accomplish using the software.

Personas guide design decisions by ensuring that the team is always thinking about the users' perspectives and needs.

User Stories

User stories provide a simple, focused description of a specific interaction from the user's point of view. They typically employ the following format.

> *As a [user type], I want to [perform a task] so that [I can achieve a goal].*

This is an example: *As a project manager, I want to generate reports so that I can track the progress of my team.*

User stories break down complex features into individual tasks that address user needs. They are an integral part of agile development processes, helping to prioritize work based on user impact.

Specific examples make the concept of personas and user stories much clearer! Let's go over a few, building on the information you provided.

Example 1: E-commerce Platform

Persona

- **Name**: Sarah Jones (a representative name only)

- **Demographic information**: 32 years old, marketing manager, lives in a suburban area, has a young child.

- **Goals and motivations**: Wants to quickly find and purchase high-quality, eco-friendly children's clothing and toys. Appreciates personalized recommendations and a seamless checkout experience.

185

- **Frustrations and pain points**: Dislikes cluttered websites, slow loading times, and complicated return processes. Finds it difficult to discover new brands and products that align with her values.

- **Technical proficiency**: Comfortable using mobile apps and online shopping platforms. Prefers intuitive interfaces and clear navigation.

- **Use-case scenarios**: Sarah uses the e-commerce platform on her smartphone during her commute to browse new arrivals, add items to her wishlist, and complete purchases using a saved payment method. She also appreciates receiving personalized product recommendations based on her past purchases and browsing history.

User Stories

- "As a busy parent, I want to be able to filter products by age range and category so that I can quickly find what I need for my child."

- "As a conscious consumer, I want to see clear information about the sustainability and ethical sourcing of products so that I can make informed purchasing decisions."

- "As a frequent shopper, I want to save my preferred payment method and delivery address so that I can check out quickly and easily."

Example 2: Project Management Software

Persona

- **Name**: David Lee (a representative name only)

- **Demographic information**: 45 years old, project manager in a software development company, works remotely and leads a team of ten developers.

- **Goals and motivations**: Needs to efficiently manage multiple projects, track progress, collaborate with his team, and report to stakeholders. Values clear visualizations of project timelines and resource allocation.

- **Frustrations and pain points**: Finds it challenging to keep track of all tasks and deadlines across different projects. Struggles with inefficient communication and collaboration tools.

- **Technical proficiency**: Highly proficient with technology and various project management methodologies.

- **Use-case scenarios**: David uses the project management software to create project plans, assign tasks, track progress using Gantt charts, communicate with his team through built-in messaging features, and generate reports for stakeholders.

User Stories

- "As a project manager, I want to create custom dashboards that show the status of all my projects at a glance so that I can easily identify potential roadblocks."

- "As a team lead, I want to be able to share files and provide feedback directly within the platform so that we can maintain a centralized and organized workspace."

- "As someone who reports to stakeholders, I want to generate professional-looking reports with customizable data visualizations so that I can effectively communicate project progress and performance."

These examples illustrate how personas and user stories can provide valuable context and focus for design and development teams. By keeping the needs and goals of specific user types in mind, teams can create software that is truly user-centric and effective.

Prototyping and Wireframing

Prototyping and wireframing are key components of the iterative design process, allowing teams to visualize and test design ideas before they are fully developed. This reduces the risk of investing time and resources into features that may not meet user needs.

Wireframes

Wireframes are low-fidelity representations of a design, focusing on structure and layout without distracting details like colors, fonts, or images. They serve as blueprints for the design, helping teams explore different layouts and content placement.

- **Purpose**: Wireframes clarify the spatial relationships between elements on the page and outline the functionality of the interface.

- **Focus on content hierarchy**: By focusing on layout rather than aesthetics, wireframes help establish content priorities and flow, ensuring that important elements stand out.

- **Tools**

  - **Balsamiq**: Known for its hand-drawn aesthetic, Balsamiq encourages rapid wireframing and ideation.

  - **Sketch**: A vector-based design tool popular for web and mobile app wireframing, offering a wide range of user interface elements and templates.

  - **Figma**: A collaborative design platform with robust wireframing capabilities, allowing for real-time collaboration and easy sharing.

- **Example**: Imagine designing a landing page for a new product. A wireframe would outline the placement of the headline, product image, call-to-action buttons, and key features, ensuring a clear and logical flow for the user.

Prototypes

Prototypes are interactive models of the software that simulate user interactions with a system. They range from low-fidelity (basic sketches linked together) to high-fidelity (interactive mockups that resemble the final product).

- **Purpose**: Prototypes allow for early testing and validation of design ideas with users, enabling the team to gather feedback before the development phase.

- **Interactive**: Unlike wireframes, prototypes allow users to experience navigation, functionality, and interaction models.

- **Tools**

  - **Figma** excels at prototyping, enabling the creation of interactive mockups with transitions and animations.

  - **Adobe XD** is a powerful prototyping tool with advanced features for creating realistic interactions and animations.

  - **InVision Studio** is a comprehensive design and prototyping platform with robust features for creating interactive experiences.

- **Example**: A prototype of a mobile banking app would allow users to simulate actions like logging in, viewing account balances, transferring funds, and paying bills. This helps identify any potential usability issues or areas for improvement in the user experience.

Both wireframes and prototypes are essential tools in user-centric design, helping to refine ideas and validate concepts before committing to full development.

User Experience and User Interface Design

Although user experience (UX) and user interface (UI) design are often discussed together, they represent different aspects of the design process. Both are essential for creating user-friendly software, but they focus on different elements of the user experience.

User Experience Design

UX design encompasses the entire journey a user takes when interacting with a product. It's about more than just usability; it's about ensuring that every interaction a user has with the product is meaningful and satisfying.

- **Focus on user flow**: UX design ensures that users can easily move through the system and complete their tasks without confusion or frustration.

- **Interaction design**: UX designers focus on how users interact with various elements (e.g., buttons, forms) and ensure that those interactions are smooth and intuitive.

- **Information architecture (IA)**: A well-structured IA helps users navigate the software more effectively by organizing content and functionality in a way that aligns with users' mental models.

User Interface Design

UI design focuses on the visual and interactive elements of the product, such as buttons, icons, typography, and color schemes. While UX design focuses on usability and user flows, UI design enhances the aesthetic and interactive appeal of the product.

- **Consistency**: Consistent UI elements help users predict behavior and navigate more easily.

- **Aesthetic design**: A visually appealing interface contributes to user satisfaction and helps build trust in the product.

- **Responsive design**: UI design must adapt to different devices, screen sizes, and resolutions to provide a seamless experience across platforms.

191

By integrating UX and UI design, developers ensure that their software is not only functional but also visually engaging and easy to use.

Usability Testing: Types and Strategies

Usability testing is a critical component of user-centric design, allowing teams to validate whether their software is intuitive and easy to use. Testing is conducted with real users to identify pain points, errors, and opportunities for improvement.

Types of Usability Testing

Moderated Usability Testing

In moderated sessions, a facilitator guides users through tasks while observing their actions and asking follow-up questions. This allows for in-depth feedback and real-time clarification.

Pros	Cons
Rich qualitative data: Provides in-depth insights into user behavior, motivations, and pain points through direct observation and interaction.	**Resource intensive**: Requires trained moderators and dedicated time for each participant.
Flexibility: Allows the moderator to adapt the test in real-time, exploring unexpected issues or delving deeper into interesting observations.	**Potential for bias**: The moderator's presence and interaction can influence user behavior, potentially leading to biased results.
Identify root causes: Facilitators can ask follow-up questions to understand the "why" behind user actions and identify the root causes of usability problems.	**Smaller sample size**: Moderated testing is typically conducted with fewer participants due to time and resource constraints.

Unmoderated Usability Testing

Users complete tasks on their own without a facilitator. This method is more scalable and cost-effective, but it lacks the in-depth feedback that moderated testing provides.

Pros	Cons
Scalability and cost-effectiveness: Can be conducted with larger sample sizes and requires fewer resources compared to moderated testing. **Reduced bias**: Eliminates the potential for moderator bias as users interact with the software independently. **Faster results**: Data collection is typically faster as users complete tests on their own schedule.	**Limited qualitative data**: Provides less in-depth insights into user motivations and thought processes. **Lack of flexibility**: Cannot adapt the test in real-time to explore unexpected issues or clarify user actions. **Technical difficulties**: Users may encounter technical difficulties during the test without a moderator to assist them.

Remote Usability Testing

Remote testing allows users to interact with the software from their own environment, providing insights into how it performs in real-world settings.

Pros	Cons
Real-world context: Captures user behavior in their natural environment, providing insights into how the software is used in real-world settings.	**Limited control**: Researchers have less control over the testing environment and may not be able to observe user behavior directly.
Geographic reach: Allows for testing with users across different locations and time zones.	**Technical challenges**: Potential for technical difficulties due to variations in user hardware, software, and Internet connectivity.
Convenience: Provides flexibility for both users and researchers, as tests can be conducted at the user's convenience.	**Reduced interaction**: May limit the opportunity for real-time interaction and follow-up questions, especially in unmoderated remote testing.

A/B Testing

This method involves presenting two versions of a feature or interface to different user groups to determine which performs better. A/B testing is especially useful for refining UI elements or content placement.

Pros	Cons
Data-driven optimization: Provides clear, quantitative data on which design performs better based on user interactions.	**Limited scope**: Primarily focuses on quantitative data and may not provide in-depth qualitative insights into user preferences.
Targeted improvements: Allows for focused testing of specific UI elements, content variations, or design choices.	**Potential for false positives**: Results can be influenced by external factors, leading to inaccurate conclusions if not carefully analyzed.
Easy to implement: Relatively easy to set up and run A/B tests using various online tools.	**Requires sufficient traffic**: Needs a significant number of users to achieve statistically significant results.

Card Sorting

Card sorting is a technique to evaluate the information architecture of a product. Users organize content into categories, which helps designers understand how users naturally group information.

Pros	Cons
Understanding user mental models: Provides insights into how users naturally categorize and organize information. **Improve information architecture**: Helps create intuitive navigation and information structures that align with user expectations. **Simple and cost-effective**: Can be conducted with relatively small groups of users and minimal resources.	**Limited scope**: Focuses specifically on information architecture and may not provide insights into other usability aspects. **Subjectivity**: Results can be subjective and may vary depending on the participants and their individual perspectives. **Requires careful analysis**: Needs careful interpretation of results to identify patterns and draw meaningful conclusions.

Key Metrics for Usability Testing

- **Task success rate**: The percentage of users who can complete a given task successfully.

- **Time on task**: How long it takes users to complete a task, indicating efficiency.

- **Error rate**: The number of errors users encounter while performing tasks.

- **Satisfaction scores**: User feedback on how satisfied they were with the experience.

By conducting usability testing at various stages of development, teams can make informed decisions about design improvements and ensure that the product is aligned with user needs.

Accessibility in Software Design

Accessibility is crucial for creating inclusive software that everyone, including people with disabilities, can use. It's not just about complying with legal requirements but also about creating a better user experience for all. Let's break down key accessibility considerations, along with tools and legal implications.

- **Keyboard navigation**: Ensure all functionality can be accessed using only the keyboard. This is essential for users who cannot use a mouse due to motor impairments.

  Example: Allow users to navigate through menus, forms, and interactive elements using the Tab key and activate them with the Enter key.

- **Screen readers**: Make sure content is compatible with screen readers, which convert text to speech or Braille for users with visual impairments.

  Example: Use semantic HTML elements (headings, lists, etc.) to structure content and provide alternative text descriptions for images.

- **Text contrast and size**: Use sufficient color contrast between text and background and allow users to adjust font sizes to ensure readability.

Example: Use a color contrast checker to ensure sufficient contrast ratios and provide options for users to increase font sizes.

- **Accessible forms**: Forms should have clear labels, and error messages should be accessible to screen readers and other assistive technologies.

Example: Use ARIA attributes to provide additional context and instructions for screen reader users and ensure error messages are programmatically associated with the relevant form fields.

- **Alt text for images**: Provide descriptive alternative text for images to convey their meaning to users who cannot see them.

Example: For an image of a product, the alt text might be "A person wearing a blue t-shirt with our company logo."

The following describes some of the available accessibility tools.

- **Automated Testing Tools**

  - **WAVE Web Accessibility Evaluation Tool**: A browser extension and online service that identifies accessibility errors and provides recommendations for improvement.

  - **Lighthouse**: An open-source tool integrated into Google Chrome that audits web pages for accessibility, performance, and other best practices.

  - **axe DevTools**: A browser extension that helps developers identify and fix accessibility issues in their code.

- **Manual Testing**

  - **Screen readers**: Use screen readers like NVDA (free) or JAWS (commercial) to experience the website as a visually impaired user would.

  - **Keyboard-only navigation**: Test the website using only the keyboard to ensure all functionality is accessible.

  - **Color contrast checkers**: Use tools like WebAIM's Color Contrast Checker to verify sufficient color contrast.

The following are some of the legal considerations.

- **Americans with Disabilities Act (ADA)**: In the United States, the ADA prohibits discrimination based on disability, including in the digital realm. Websites and software must be accessible to people with disabilities.

- **Similar international laws**: Many countries have similar legislation, such as the Accessibility for Ontarians with Disabilities Act (AODA) in Canada and the Equality Act 2010 in the United Kingdom.

- **Consequences of non-compliance**: Organizations can face legal action, fines, and reputational damage for failing to provide accessible digital experiences.

By following accessibility best practices and utilizing the available tools, developers can create inclusive software that benefits everyone. Businesses need to understand the legal and ethical implications of accessibility and prioritize it in their software development processes.

Iterative Design: The Role of Feedback

In user-centric design, feedback is not only valuable but necessary for success. Iterative design is an approach that emphasizes continuous refinement of the product based on user feedback. This process allows for ongoing adjustments to ensure that the final product meets the needs of its users.

The Iterative Design Process

- **Prototype**: Create a low-fidelity or high-fidelity prototype based on user research.

- **Test**: Conduct usability testing or gather user feedback on the prototype.

- **Analyze**: Review the feedback to identify areas for improvement.

- **Refine**: Make adjustments to the design based on the feedback.

- **Repeat**: Continue testing and refining until the product meets usability goals.

The Importance of Feedback Loops

- **User-centered adjustments**: Feedback from real users ensures that design decisions are grounded in user needs rather than assumptions.

- **Early problem detection**: Iterative testing allows teams to identify and resolve usability issues before they become costly to fix.

- **Continuous improvement**: The iterative design process encourages incremental improvements, ensuring that the product evolves alongside user expectations and market demands.

By embracing iterative design, teams can create software that is highly responsive to user needs, reducing the likelihood of major usability issues in the final product.

Cognitive Load and Design Simplicity

Cognitive load refers to the amount of mental effort required to process information and complete tasks. High cognitive load can overwhelm users, leading to confusion, frustration, and mistakes. Simplifying the design is crucial to minimizing cognitive load and improving usability.

Types of Cognitive Load

- **Intrinsic load**: The inherent difficulty of the task itself. For example, learning a new programming language has a higher intrinsic load than using a familiar tool.

- **Extraneous load**: The cognitive effort required due to poorly designed interfaces or unnecessary complexity. This is the type of load that designers should aim to reduce.

- **Germane load**: The effort required for learning and understanding. In software, this relates to how much users must learn to use the system effectively.

Strategies to Reduce Cognitive Load

- **Simplify navigation.** Ensure that users can easily find their way through the system without encountering dead ends or confusing paths.

- **Minimize choices.** Too many options can overwhelm users. Simplify decision-making by providing clear choices and reducing the number of actions on each screen.

- **Provide visual cues.** Use consistent icons, labels, and colors to guide users through tasks and reinforce actions.

- **Focus on recognition over recall.** Design systems that allow users to recognize familiar patterns and actions rather than relying on memory to recall instructions or commands.

By reducing cognitive load, designers make it easier for users to interact with the software, leading to faster task completion, fewer errors, and a more positive experience overall.

Navigation and Information Architecture

Effective navigation and information architecture are essential for a positive user experience, especially in complex software systems. Navigation refers to how users move through the software, while IA is the organization and labeling of information to make it easier for users to find what they need.

Key Principles of Navigation Design

- **Consistency**: Maintain consistent navigation elements across all screens. Users should not need to relearn how to navigate each time they move to a new section of the software.

- **Visibility**: Navigation elements should be easily visible and accessible. Hidden menus or confusing layouts can frustrate users and lead to higher abandonment rates.

- **Contextual feedback**: Provide feedback on the user's location within the system. Highlight the current page in the navigation bar and use breadcrumb trails to show the user's path.

- **Hierarchical structure**: Organize content into a clear hierarchy, with primary navigation for major sections and secondary navigation for related sub-sections.

Information Architecture Best Practices

A well-designed navigation and IA system allows users to easily explore the software and complete tasks with minimal confusion or effort.

- **User-centered labeling**: Use language that resonates with users, avoiding jargon or overly technical terms. Labels should clearly indicate the purpose of each link or button.

- **Search functionality**: Ensure that users can easily search for specific content, especially in systems with large datasets or complex structures.

- **Content grouping**: Group related content together logically, based on how users expect to find information.

Tools and Methodologies for Usability

Various tools and methodologies support usability in software design, helping teams build, test, and refine their products.

Usability Tools

- **Figma** is a powerful design tool for creating wireframes, prototypes, and high-fidelity designs. It allows real-time collaboration among team members.

- **InVision** is a prototyping tool that enables designers to create interactive, clickable prototypes. It also facilitates feedback collection from stakeholders.

- **Hotjar** is a user analytics tool that tracks user behavior through heatmaps, session recordings, and conversion funnels. It helps teams understand where users struggle to navigate or interact with the software.

- **Google Analytics** provides insights into user behavior, showing how users engage with different parts of the system. It can help identify areas of the software that need improvement.

Usability Methodologies

- **Lean UX** is a methodology that emphasizes rapid prototyping and user feedback over heavy documentation. The goal is to iterate quickly based on real user insights.

- **Agile UX** integrates UX design into agile development processes, ensuring that user feedback is incorporated into each sprint cycle.

- **Design thinking** is a human-centered design approach that focuses on understanding the user's needs, brainstorming solutions, prototyping, and testing ideas.

These tools and methodologies help design teams stay user-focused throughout the development process, ensuring that usability remains a priority from concept to launch.

Usability in Mobile and Web Applications

The rise of mobile and web applications has introduced unique challenges and considerations for usability. Given the different screen sizes, interaction methods, and user contexts, designing for usability in mobile and web apps requires a specialized approach.

Mobile Usability Considerations

- **Touch interactions**: Ensure that buttons and other interactive elements are large enough to be tapped easily on small screens. Gestures like swiping or pinching should be intuitive and aligned with common mobile conventions.

- **Responsive design**: Mobile apps must adapt to various screen sizes and orientations. Responsive design ensures that the interface remains usable whether the user is on a phone, tablet, or desktop.

- **Minimalist design**: Given the limited screen space on mobile devices, prioritize key tasks and keep the interface clean and uncluttered.

- **Offline access**: Many mobile users need to access apps without a reliable Internet connection. Consider designing for offline functionality where possible.

Web Application Usability Considerations

- **Loading time**: Users expect web applications to load quickly. Optimize page performance to reduce loading times and prevent frustration.

- **Cross-browser compatibility**: Ensure that the web application functions consistently across different browsers and devices, including desktops, tablets, and mobile phones.

- **Progressive disclosure**: Only show the most important information and options up front. Additional details can be revealed when needed, reducing cognitive load.

- **Accessible forms**: Forms are a key element in web applications. Ensure that forms are easy to fill out, error messages are clear, and fields are properly labeled.

Adaptive and Responsive Design

Both mobile and web applications should prioritize usability to meet user expectations for efficiency, reliability, and ease of use.

- **Responsive design**: Automatically adjusts the layout and elements based on the screen size. This approach ensures that users have a seamless experience across devices.

- **Adaptive design**: Provides different layouts based on specific screen sizes or breakpoints. It delivers more tailored experiences but requires more development effort.

Measuring and Improving Usability

To continuously improve the user experience, it's essential to measure usability and use the data to guide design decisions.

Key Usability Metrics

- **Task success rate**: The percentage of users who successfully complete a given task without errors. A high success rate indicates good usability.

- **Time on task**: The amount of time it takes users to complete a task. Shorter completion times generally reflect more efficient design, provided accuracy is not compromised.

- **Error rate**: The number of errors users make when interacting with the software. Reducing the error rate is key to improving user satisfaction.

- **Net Promoter Score (NPS)**: A measure of user satisfaction based on how likely users are to recommend the product to others.

- **System Usability Scale (SUS)**: A standardized questionnaire used to evaluate the overall usability of a system. It provides a single usability score based on a user's responses to a series of questions.

Techniques for Improving Usability

- **Heuristic evaluation**: Usability experts review the software against established usability principles (heuristics) to identify potential issues.

- **A/B testing**: Experiment with different design elements and measure the impact on user behavior. This is especially useful for optimizing UI components and workflows.

- **Continuous user feedback**: Implement feedback mechanisms within the software, such as surveys or in-app chat, to gather ongoing insights into user satisfaction and pain points.

By measuring usability regularly and making data-driven decisions, teams can ensure that the software continues to meet user needs and deliver an exceptional experience.

Challenges in Implementing User-Centric Design

While user-centric design offers significant benefits, implementing it in practice comes with challenges that teams must navigate.

Common Challenges

- **Balancing user needs with business goals**: While users may prefer certain features or workflows, these must align with the company's business objectives. Teams must find a balance between usability and technical or business constraints.

- **Resource limitations**: User research, usability testing, and iterative design processes can be resource-intensive. Limited budgets or time constraints may pressure teams to skip important steps, leading to suboptimal user experiences.

- **Stakeholder misalignment**: Stakeholders may have differing opinions on what constitutes good design, leading to conflicts in decision-making. It's crucial to keep user data and usability metrics at the forefront of discussions to drive alignment.

- **Resistance to change**: Established companies or teams may resist adopting user-centric practices, particularly if they are accustomed to a more traditional development process. Educating stakeholders on the long-term benefits of user-centric design is key to overcoming this resistance.

Strategies for Overcoming Challenges

- **Start small**: Implement user-centric practices incrementally, such as starting with user testing or personas, before scaling to a full UCD process.

- **Educate stakeholders**: Share usability testing results and user feedback with stakeholders to demonstrate the value of a user-centered approach.

- **Prioritize iterative design**: Focus on making continuous improvements rather than aiming for perfection in the first release. This allows teams to refine the product over time based on user feedback.

Despite the challenges, adopting user-centric design methodologies leads to better products, happier users, and, ultimately, greater business success.

Conclusion

User-centric software strategies are essential for creating software that is intuitive, efficient, and satisfying for users. By focusing on the user experience from the beginning and involving users throughout the design process, teams can ensure that their products meet real user needs and deliver a delightful experience.

This chapter explored the principles of user-centered design, the importance of understanding user needs, and the value of iterative feedback. It also discussed key tools and methodologies that help ensure usability is prioritized during development, from wireframing and prototyping to usability testing and accessibility considerations.

In conclusion, adopting a user-centric approach is not just a technical decision; it is a strategic one. It requires a commitment to understanding the user's perspective, a willingness to iterate and improve, and a focus on delivering long-term value through superior user experiences. The reward for this commitment is software that not only functions well but also fosters loyalty, satisfaction, and engagement from users.

PART III

Governance

CHAPTER 8

Ensuring Data Quality and Integrity

Data is one of the most valuable assets in modern organizations. With data driving everything from daily operations to strategic decision-making, ensuring its quality and integrity is crucial. Poor-quality data can lead to costly errors, inefficiencies, and missed opportunities, while compromised data integrity can result in mistrust, legal complications, and operational breakdowns. As organizations generate, collect, and process vast amounts of data, maintaining its quality and ensuring its integrity has become both more challenging and more critical.

Ensuring data quality and integrity means establishing trust in the data being used, which involves maintaining high standards across accuracy, consistency, completeness, and reliability. Achieving these objectives requires a comprehensive approach that includes data validation, cleaning, governance, and continuous monitoring. Additionally, the role of emerging technologies such as artificial intelligence and machine learning in detecting anomalies and automating data quality processes is rapidly growing.

This chapter explores the principles, tools, and techniques for ensuring data quality and integrity, providing a deep dive into best practices, challenges, and real-world applications.

P. Gujar, *Data Usability in the Enterprise*, https://doi.org/10.1007/979-8-8688-1183-8_8

Defining Data Quality

At its core, data quality refers to the degree to which data meets the standards required for it to serve its intended purpose effectively. This definition varies by context, but universally, it involves assessing data on multiple dimensions, such as accuracy, completeness, and timeliness. High-quality data is essential for analytical processes, operational efficiency, compliance with regulations, and strategic decision-making.

Why is Data Quality Important?

The value of data lies not just in its quantity but in its quality. Inaccurate or incomplete data can lead to wrong conclusions, which can, in turn, lead to incorrect business decisions.

- In healthcare, poor data quality can lead to inaccurate diagnoses, ineffective treatments, or compromised patient safety.

- In finance, errors in transactional or customer data can lead to regulatory penalties, financial loss, and reputational damage.

- In supply chain management, data inaccuracies can disrupt logistics, leading to stock shortages or overstocking.

High-quality data ensures that business processes run smoothly, increases operational efficiency, and supports informed decision-making across all levels of an organization.

Components of Data Quality

There are many ways to define and measure data quality. The following are some common components.

- **Fit for purpose**: Data must meet the needs of its intended use, whether for decision-making, analysis, or operations.

- **Reliability**: Data should be trustworthy and dependable, meaning that users can rely on it to perform the expected operations accurately and consistently.

- **Relevance**: Data must be relevant to the objectives or decisions being made. Unrelated or outdated data diminishes the quality and utility of the dataset.

A comprehensive approach to managing data quality ensures that organizations can maintain confidence in their data, leading to better outcomes in both the short and long term.

Key Dimensions of Data Quality

To ensure comprehensive data quality, organizations must evaluate multiple dimensions that determine how well the data serves its purpose. Each dimension addresses a specific characteristic of data that contributes to its overall quality.

Accuracy

Accuracy is the most critical dimension of data quality, ensuring that the data correctly represents the real-world objects or events it is meant to model. For example, if an employee's name or salary is recorded inaccurately, it can lead to payroll errors or legal issues. Accuracy also depends on the precision of the measurement systems and how well they capture real-world values.

Completeness

Completeness refers to whether all necessary data points are present. Missing data can lead to incomplete analyses or incorrect conclusions. For example, if a customer record is missing key fields such as contact information, it can impede follow-ups or marketing campaigns. To assess completeness, organizations typically check for null or empty fields, missing records, or gaps in time series data.

Consistency

Consistency ensures that the same piece of data is uniform across different systems, databases, or datasets. Inconsistencies often arise when data is duplicated in multiple locations without proper synchronization. For example, an order system might show a different shipping address than the CRM system, leading to confusion and errors in processing customer orders. Consistency checks help prevent this by ensuring that data is uniform across systems and does not contain conflicting values.

Timeliness

Timeliness refers to how current the data is and whether it is available when needed. In many industries, such as financial trading or healthcare, timely data is critical. For example, stock market data that is just seconds out of date can result in missed trading opportunities. Timeliness also involves processing data efficiently so that decision-makers can act on it quickly.

Validity

Validity ensures that the data follows the required formats, rules, or constraints. For instance, a field for recording dates must follow a specific format, such as YYYY-MM-DD. Additionally, validity ensures that only allowable values are entered, such as checking whether a field for recording age contains only positive numbers.

Uniqueness

Uniqueness ensures that there are no duplicate records in a dataset. Duplicate data can arise when data entry systems do not prevent multiple entries for the same entity, such as a customer or product. Duplication can lead to inefficiencies and incorrect reporting, such as inflated counts or duplicated efforts in marketing campaigns.

Integrity

Data integrity is another critical dimension, encompassing accuracy, consistency, and reliability. It is typically ensured through mechanisms like referential integrity in databases, which make sure that relationships between different tables or datasets remain valid and intact over time.

Evaluating data quality across these dimensions provides a more comprehensive understanding of the strengths and weaknesses of a dataset. Organizations must implement strategies to maintain quality across all dimensions so that their data remains valuable.

Understanding Data Integrity

Data integrity refers to the trustworthiness of data throughout its lifecycle—whether it's during creation, storage, transmission, or modification. Data integrity ensures that data remains accurate, consistent, and unaltered unless through authorized processes.

Types of Data Integrity

- **Entity integrity** ensures that each entity (or row) in a database has a unique identifier, typically a primary key. This helps prevent duplication of records and maintains uniqueness, ensuring data is not lost or overwritten erroneously.

  Example: In a customer database, the primary key might be the customer's ID, ensuring that each record is unique.

- **Referential integrity** ensures that relationships between different tables in a database are valid. For example, if there's a relationship between customer records and orders, referential integrity ensures that each order references a valid customer ID. Without referential integrity, it would be possible to have orders linked to non-existent customers, leading to errors in reporting and processing.

- **Domain integrity** ensures that the values entered into a specific field conform to predefined rules. For example, a field for recording a person's age should only accept positive integers, while a "gender" field might only accept specific options such as Male, Female, or Other.

- **User-defined integrity rules** are defined by an organization to meet specific business needs. These user-defined rules ensure that data adheres to business logic, such as making sure that an employee's end date cannot precede their start date.

The Role of Integrity Constraints

Integrity constraints are rules applied to ensure that data adheres to its defined standards. The following are common integrity constraints.

- **Unique constraints** ensure that a particular field (e.g., email address) contains unique values within a dataset.

- **Foreign key constraints** enforce relationships between tables, such as ensuring that a sales order references an existing customer.

- **Check constraints** validate that data entered into a field meets specified criteria, such as ensuring that numerical values fall within a certain range.

Maintaining data integrity helps prevent data corruption, ensures data accuracy, and supports regulatory compliance. Failing to ensure integrity can lead to data that is not trusted, rendering it ineffective for decision-making or analysis.

Causes of Data Quality and Integrity Issues

Data quality and integrity issues can arise from a variety of sources, both human and technical. Understanding these causes is key to implementing effective strategies for prevention and remediation.

Human Error

Human error is one of the most common causes of data quality issues. This includes manual data entry mistakes, such as inputting incorrect values, omitting critical data, or entering duplicate records. Inconsistent or unclear data entry guidelines can exacerbate these problems, leading to discrepancies in how data is recorded across the organization.

Data Decay

Over time, data can "decay" or become outdated, especially in fast-moving environments where customer details, market conditions, or inventory levels frequently change. Data that was once accurate can become incomplete or incorrect if it is not regularly updated. For example, customer addresses or phone numbers may change over time, leading to incomplete or inaccurate contact information.

System Integration and Synchronization Issues

In many organizations, data is stored across multiple systems or databases that must communicate and share information. Integration issues can lead to synchronization problems, where data in one system does not match data in another. For instance, if an e-commerce system and a CRM system are not properly synchronized, a customer order may not appear in the CRM system, leading to incomplete order records and a poor customer experience.

Data Migration

Migrating data from one system to another often introduces errors, especially when different systems use different data formats or have varying data standards. Data might be lost, duplicated, or corrupted during migration if it's not handled carefully. Testing and validation are critical steps in the migration process to ensure that data remains intact.

Poor Data Governance

Without strong data governance practices, data quality and integrity issues can proliferate. Poor governance leads to inconsistent data entry practices, lack of accountability, and unclear ownership of data quality responsibilities. Additionally, without governance, data standards and policies may not be enforced, resulting in varying levels of quality across the organization.

Technical Failures

Hardware malfunctions, software bugs, or network issues can lead to data corruption or loss. For example, power outages during data transmission or database operations can result in incomplete or corrupted transactions. Backups and redundant systems can mitigate these risks, but without proper safeguards, technical failures can compromise data quality and integrity.

Inadequate Validation and Testing

If data validation checks are insufficient, errors may go unnoticed. Organizations must implement robust validation processes, including automated checks at data entry points and manual reviews of critical datasets. Regular testing of systems and processes ensures that data validation remains effective as the organization grows or changes.

Understanding these causes allows organizations to target the root of their data quality and integrity challenges, leading to more effective prevention and remediation efforts.

Best Practices for Ensuring Data Quality

Ensuring data quality requires a multifaceted approach that includes policies, procedures, and technology. The following best practices can help organizations improve and maintain high data quality.

Establish Clear Data Standards

Data standards provide a framework for how data should be collected, formatted, and stored. These standards should be defined based on the needs of the organization and the specific use cases for the data. For example, the format for dates (e.g., MM-DD-YYYY) should be consistent across systems. Additionally, standards for naming conventions, permissible data ranges, and mandatory fields must be defined.

Having clear standards reduces ambiguity, minimizes the likelihood of errors, and ensures that data is collected in a consistent and structured manner. Documentation of these standards is critical for communicating them to all stakeholders and ensuring compliance.

Implement Rigorous Data Validation

Data validation checks are essential for catching errors before they enter the system. Validation should occur at the point of data entry, ensuring that data adheres to the predefined standards. Automated validation can check for errors like the following.

- **Field validation**: Ensuring that each field contains valid data (e.g., a phone number field should not contain letters).

- **Range checks**: Ensuring that numerical values fall within an acceptable range (e.g., an age field should contain values between 0 and 120).

- **Cross-field validation**: Ensuring that related fields are logically consistent (e.g., the shipping date cannot be before the order date).

Validation rules should be revisited regularly to ensure they continue to meet the needs of the organization.

Encourage a Data Quality Culture

Data quality is not just a technical issue—it requires a cultural shift across the organization. Employees at all levels must understand the importance of data quality and be trained on the best practices for data entry, maintenance, and review. Encourage a culture of accountability where everyone is responsible for maintaining data quality in their respective areas.

Data Stewardship and Accountability

Appoint data stewards who are responsible for overseeing data quality in specific domains. These stewards ensure that data policies are followed, resolve data quality issues, and promote best practices. Data stewardship ensures that there is clear ownership of data quality responsibilities, reducing the risk of errors being overlooked or ignored.

Regular Data Audits and Cleansing

Data should be regularly audited to identify and resolve any quality issues, such as duplicates, missing data, or outdated information. Data cleansing processes help ensure that the dataset remains accurate and consistent over time. Automated data profiling tools can assist in identifying these issues, but manual reviews may also be necessary for critical data fields.

Implement Data Governance Frameworks

Effective data governance is crucial for maintaining data quality over time. A well-defined data governance framework includes policies, standards, and roles that support consistent data management practices. This includes defining data ownership, enforcing compliance with data standards, and establishing procedures for handling data quality issues.

By implementing these best practices, organizations can reduce the likelihood of data quality issues and create a more robust, reliable data environment.

Data Quality Management Frameworks

Data quality management frameworks provide structured approaches for ensuring that data quality is maintained across an organization. These frameworks outline processes for data governance, validation, monitoring, and improvement.

Total Data Quality Management

Total Data Quality Management (TDQM) is a comprehensive framework focused on continuous improvement of data quality. TDQM views data quality as a cycle that involves four key steps.

- **Define**: Establish what constitutes high-quality data by defining quality criteria and metrics. These definitions should be aligned with the organization's strategic goals and the needs of its stakeholders.

- **Measure**: Regularly assess the current state of data quality by comparing actual data to the defined metrics. This includes assessing accuracy, completeness, and timeliness.

- **Analyze**: Identify the root causes of any data quality issues and determine the impact of poor-quality data on the organization. Analyzing trends and patterns helps pinpoint systemic issues.

- **Improve**: Implement corrective actions to address identified data quality problems. This may involve revising data collection processes, updating validation rules, or introducing new technology.

TDQM emphasizes the importance of continuous monitoring and improvement, ensuring that data quality efforts evolve along with the organization's needs.

Data Quality Assessment

The Data Quality Assessment (DQA) framework involves a detailed evaluation of a dataset to determine its strengths and weaknesses. The assessment typically covers multiple dimensions of data quality, including accuracy, completeness, and consistency.

DQA typically follows these steps.

1. **Initial assessment**: Perform a baseline assessment to identify existing data quality issues. This may involve profiling the data, running validation checks, and gathering feedback from stakeholders.

2. **Gap analysis**: Compare the current state of the data to the desired state, identifying gaps and areas for improvement.

3. **Action plan**: Develop a strategy for addressing data quality gaps. This may include process changes, data cleansing, or technology upgrades.

4. **Ongoing monitoring**: Continuously monitor data quality to ensure that improvements are sustained over time.

DQA is particularly useful in organizations where data quality is critical to compliance, reporting, or decision-making.

ISO 8000-61: Data Quality Management Standard

The ISO 8000-61 standard provides guidelines for organizations to manage data quality effectively. It emphasizes the importance of defining data requirements, establishing processes to ensure quality, and regularly monitoring and measuring performance.

ISO 8000-61 provides a formalized approach to data quality management, helping organizations comply with industry regulations and achieve consistent, high-quality data across all business functions.

By adopting a data quality management framework, organizations can implement a structured approach to maintaining data quality over time, ensuring that data remains a valuable asset.

Data Validation Techniques

Data validation is a critical step in ensuring data quality and integrity. Effective validation prevents errors from entering the system, ensuring that only valid, complete, and consistent data is stored and processed. Various techniques can be employed depending on the type of data and the system's requirements.

Real-Time vs. Batch Validation

Validation can be performed either in real time as data is entered into the system or in batch processes that validate data after it has been collected.

- **Real-time validation:** In this approach, data is validated as it is entered, providing immediate feedback to the user. This method prevents invalid data from being stored in the system. For example, an online form might prevent a user from submitting the form if required fields are missing or contain invalid values.

- **Batch validation:** Batch validation processes large volumes of data at predefined intervals, checking for errors after the data has been entered into the system. Batch validation is useful for large datasets, but it may not provide immediate feedback. For example, a nightly batch job might validate customer records added throughout the day.

Field-Level Validation

Field-level validation checks the data entered into specific fields for accuracy, format, and completeness. The following are examples of field-level validation.

- **Format validation:** Ensures that the data matches the expected format, such as phone numbers being entered in the correct pattern (e.g., "123-456-7890").

- **Length validation:** Ensures that the data entered into a field meets the required length, such as limiting text fields to 100 characters.

- **Required field validation:** Ensures that mandatory fields are not left blank.

Cross-Field Validation

Cross-field validation checks the relationships between fields to ensure that data is logically consistent. The following are two examples.

- **Date validation**: Ensures that an employee's hire date is not after their termination date.

- **Address validation**: Ensures that the city and state fields are compatible (e.g., "New York" should be paired with "NY").

Referential Integrity Checks

Referential integrity checks ensure that relationships between different tables or datasets are maintained. These checks are critical in relational databases where foreign keys link records across multiple tables. For example, an order record should not reference a non-existent customer ID. Enforcing referential integrity prevents orphaned records, ensuring data consistency across the system.

Business Rule Validation

Business rule validation ensures that data adheres to the organization's specific policies or business logic. For example, a retail system may have rules that prevent an order from being placed if the total order value is below a certain threshold. Business rule validation helps enforce organizational policies and prevents invalid transactions from being processed.

Post-Entry Validation

In some cases, validation occurs after data has been entered and processed. This is common in systems where data is aggregated from multiple sources, and inconsistencies or errors need to be resolved. Post-entry validation might include the following.

- **Consistency checks**: Ensuring that data remains consistent across different sources.

- **Duplicate detection**: Identifying and resolving duplicate records in large datasets.

Validation is a critical process for ensuring data quality, reducing the likelihood of errors, and maintaining the integrity of the system.

Data Cleaning and Transformation

Data cleaning (or cleansing) and transformation are essential processes for improving the quality and usability of data. Cleaning addresses errors and inconsistencies, while transformation restructures data to make it more useful for analysis or operational tasks.

The Data Cleaning Process

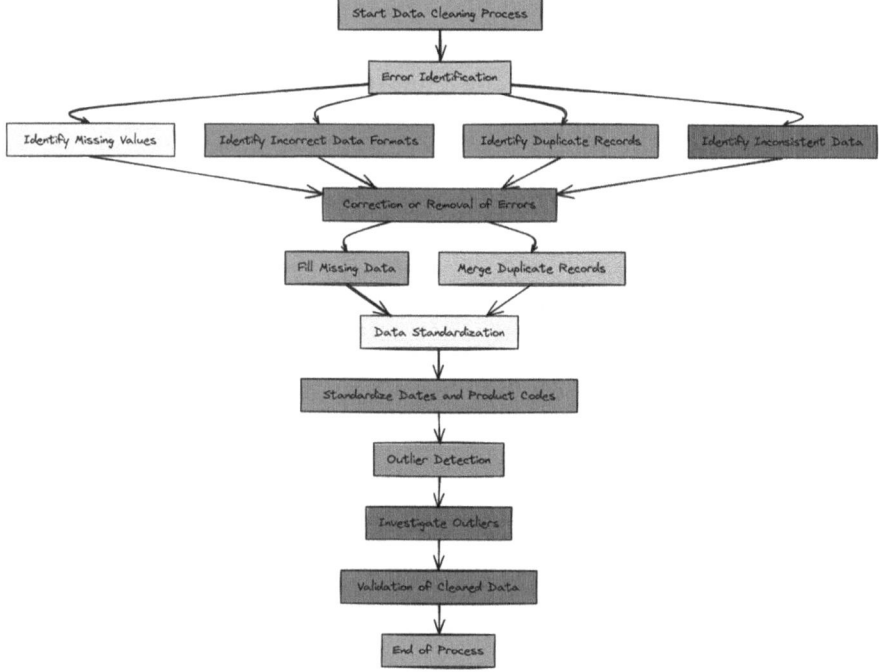

Figure 8-1. *Flowchart illustrating the data cleaning process*

The data cleaning process involves several steps designed to identify and correct errors in a dataset. The following are some key steps.

- **Error identification**: The first step is to identify the types of errors present in the data, such as

    - Missing values

    - Incorrect data formats

    - Duplicate records

    - Inconsistent data across fields or datasets

- **Correction or removal of errors**: After identifying errors, the next step is to correct them or remove faulty data from the dataset, such as

  - **Missing data** might be filled in with default values, averages, or estimations based on other available data.

  - **Duplicate records** are merged to ensure that only unique data remains.

- **Data standardization**: Standardizing data ensures that it adheres to the same format and structure across the dataset. For example, dates may be standardized to the YYYY-MM-DD format, and product codes may be formatted consistently.

- **Outlier detection**: Outliers—values that fall far outside the expected range—can skew analyses and should be investigated. Sometimes, outliers indicate errors, while other times, they represent valid but extreme cases.

- **Validation of cleaned data**: After cleaning, the data should be revalidated to ensure that the corrections did not introduce new errors. Automated tools can assist in verifying the integrity of the cleaned data.

Data Transformation

Data transformation is the process of converting data from one format or structure to another to meet the needs of specific systems or analyses. Transformation is often required when migrating data between systems, integrating data from multiple sources, or preparing data for analysis.

The following are key data transformation techniques.

- **Data type conversion**: Converting data types (e.g., changing a text field into a numerical value).

- **Data aggregation**: Summarizing data by aggregating records, such as calculating total sales for each region.

- **Normalization**: Converting data to a standard scale, such as normalizing customer ratings to a 1–10 scale.

- **Deriving new fields**: Creating new fields based on existing data, such as calculating a customer's age from their date of birth.

Both data cleaning and transformation are essential for maintaining high-quality datasets that can be trusted for analysis, reporting, and operational tasks.

Data Governance and Compliance

Data governance provides a framework for managing data effectively across an organization. It involves defining roles, policies, and processes for maintaining data quality, integrity, and security. Additionally, compliance with regulatory requirements is a critical aspect of data governance, ensuring that data is handled responsibly and in accordance with legal mandates.

Key Elements of Data Governance

Effective data governance frameworks typically include the following components.

- **Data ownership and stewardship**: Assigning ownership of data to specific individuals or teams ensures accountability. Data stewards are responsible for overseeing the quality, consistency, and integrity of data within their domain. For example, in a healthcare organization, a data steward might be responsible for ensuring the accuracy and completeness of patient records.

- **Data policies and standards**: Clearly defined policies and standards ensure that data is managed consistently across the organization. This includes defining acceptable data formats, validation rules, and access controls. For example, a financial institution might have strict policies on data retention and disposal to comply with regulatory requirements.

- **Data lifecycle management**: Governance frameworks must address the entire data lifecycle, from creation to archival. This includes defining processes for data collection, storage, usage, and eventual disposal. For example, a retail company might have a data lifecycle management process that includes collecting customer data through loyalty programs, storing it securely, using it for personalized marketing campaigns, and eventually anonymizing or deleting it according to data retention policies.

- **Data quality management:** Governance frameworks should include processes for monitoring, reporting, and improving data quality. Data quality metrics should be regularly assessed, and any issues should be addressed promptly. For example, a manufacturing company might use data quality tools to identify and correct errors in their production data, ensuring accurate inventory management and supply chain optimization.

- **Risk management and security:** Data governance frameworks must include measures to ensure data security and protect sensitive information. This involves implementing encryption, access controls, and auditing mechanisms to safeguard data. For example, a government agency might implement strict access controls and encryption to protect confidential citizen data from unauthorized access.

Regulatory Compliance

Many industries are subject to strict regulations regarding data quality, privacy, and security. Failing to comply with these regulations can result in hefty fines, legal penalties, and reputational damage. The following are key regulations.

- **General Data Protection Regulation (GDPR):** In the European Union, GDPR mandates strict requirements for the protection of personal data, including ensuring data accuracy and providing individuals with the right to access and correct their data.

- **Health Insurance Portability and Accountability Act (HIPAA):** In the United States, HIPAA regulates the handling of personal health information (PHI) and mandates strict standards for data integrity, security, and privacy.

- **Sarbanes-Oxley Act (SOX):** In the financial sector, SOX requires organizations to maintain accurate and reliable financial data, with strict controls to ensure data integrity.

By implementing strong data governance frameworks and ensuring compliance with regulatory requirements, organizations can protect their data assets, reduce risks, and maintain customer trust.

The following is a success story and a cautionary tale.

- **Success:** Companies like Amazon and Google have implemented robust data governance frameworks, leading to improved data quality, better decision-making, and increased customer trust.

- **Failure:** In 2017, Equifax suffered a massive data breach due to poor data governance practices, exposing the personal information of millions of customers and resulting in significant financial and reputational damage.

Automation and Data Quality Tools

As the volume and complexity of data grow, manually maintaining data quality becomes increasingly challenging. Automation offers a powerful solution, enabling organizations to efficiently manage data quality at scale. Automated tools can streamline validation, cleaning, and monitoring processes, freeing up human resources for higher-level tasks.

Data Quality Management Tools

Several tools are available to help organizations automate their data quality efforts.

- **Data profiling tools**: These tools analyze datasets to identify patterns, inconsistencies, and errors. They provide insights into the completeness, accuracy, and consistency of the data, enabling organizations to quickly identify and address quality issues.

- **Data cleansing tools**: Data cleansing tools automatically detect and correct common data quality problems such as duplicates, missing values, and formatting errors. These tools can be configured to apply predefined rules and standards, ensuring that data is cleaned consistently.

- **Data governance platforms**: Data governance platforms provide centralized control over data policies, lineage tracking, and quality monitoring. They help organizations enforce data standards, track changes, and ensure compliance with governance frameworks.

Benefits of Automation

Automation provides several key benefits for ensuring data quality.

- **Consistency**: Automated processes apply data quality rules consistently across the entire dataset, reducing the risk of human error.

- **Scalability**: Automation enables organizations to manage large datasets efficiently, applying validation and cleaning processes in real-time or at scheduled intervals.

- **Efficiency**: Automated tools can process vast amounts of data in a fraction of the time it would take to do manually, enabling faster resolution of data quality issues.

- **Real-time monitoring**: Automated tools can monitor data quality in real-time, alerting users to potential issues as they arise. This proactive approach helps organizations address problems before they impact operations.

By leveraging automation, organizations can ensure that data quality efforts are scalable, consistent, and efficient, even as data volumes grow.

Monitoring and Reporting Data Quality

Continuous monitoring is essential for maintaining high data quality. Monitoring involves tracking key data quality metrics, detecting anomalies, and addressing issues before they affect business operations or decision-making.

Key Data Quality Metrics

To effectively monitor data quality, organizations should track metrics that reflect the health of their data. The following are common metrics.

- **Error rate**: The percentage of records that contain errors, such as missing or invalid values. A high error rate may indicate problems with data entry processes or validation rules.

- **Completeness**: The percentage of required fields that are filled out. Incomplete data can skew analyses and reduce the accuracy of reports.

- **Timeliness**: The speed at which data is captured, processed, and made available for use. Timeliness is critical in real-time systems such as financial trading or logistics operations.

- **Duplication rate**: The percentage of records that are duplicates. Duplicate records can lead to inefficiencies, incorrect reporting, and confusion in customer-facing processes.

- **Data drift**: Changes in data patterns over time that may indicate a degradation in data quality or a shift in the underlying data-generating process.

Real-Time Monitoring and Alerts

Many data quality tools provide real-time monitoring capabilities, enabling organizations to detect data quality issues as they arise. Alerts can be configured to notify users when specific thresholds are exceeded, such as when error rates rise above a certain level or when data freshness falls below an acceptable standard.

Real-time monitoring allows organizations to respond quickly to data quality issues, minimizing the impact on operations and decision-making.

Reporting and Dashboards

Data quality dashboards provide a visual representation of key metrics, making it easy to track the health of the organization's data over time. Dashboards typically include charts, graphs, and trend analysis, allowing users to quickly identify problem areas and track improvements in data quality.

Reporting tools can generate periodic reports on data quality metrics, providing insights into trends, patterns, and recurring issues. These reports can be shared with stakeholders to demonstrate the effectiveness of data quality efforts and to identify areas for improvement.

By continuously monitoring and reporting on data quality, organizations can ensure that their data remains accurate, reliable, and fit for purpose.

Role of AI and Machine Learning in Data Quality

Artificial intelligence (AI) and machine learning (ML) technologies are transforming the way organizations manage data quality. These technologies enable the automation of complex data quality tasks, such as anomaly detection, data cleaning, and pattern recognition, making it easier to maintain high-quality data at scale.

AI for Data Profiling

AI-driven data profiling tools can automatically analyze large datasets to identify patterns, inconsistencies, and anomalies. These tools can learn from the data and adapt their analyses as new data is added, improving their ability to detect subtle or complex data quality issues over time.

For example, AI can identify data anomalies by detecting outliers that deviate from expected patterns. In a financial dataset, AI might flag transactions that fall outside the typical range for a given customer, indicating potential fraud or data entry errors.

Machine Learning for Anomaly Detection

Machine learning models can be trained to detect anomalies in data, such as outliers, unexpected trends, or deviations from normal behavior. These models can analyze large datasets in real time, providing immediate feedback when potential data quality issues are identified.

Anomaly detection models are particularly useful for identifying rare or hard-to-detect errors that traditional rule-based validation techniques might miss. For example, in an e-commerce system, machine learning might detect unusual patterns in customer orders that indicate fraudulent activity or data entry errors.

Predictive Data Quality

AI and machine learning can also be used to predict potential data quality issues before they occur. By analyzing historical data and identifying patterns, predictive models can forecast when data is likely to become inaccurate, incomplete, or inconsistent.

For example, predictive models might identify customers whose contact information is likely to become outdated based on historical trends, allowing organizations to proactively reach out for updates before data quality deteriorates.

Automated Data Cleaning

AI-powered data cleaning tools can automatically identify and correct common data quality issues, such as duplicates, missing values, and formatting errors. These tools use machine learning algorithms to learn from previous cleaning operations, improving their accuracy and efficiency over time.

For example, an AI-driven tool might automatically merge duplicate customer records by analyzing similarities in names, addresses, and transaction histories, reducing the need for manual intervention.

By leveraging AI and machine learning, organizations can improve the accuracy, consistency, and reliability of their data while reducing the time and effort required to maintain data quality.

While AI and machine learning offer powerful tools for enhancing data quality, it's important to be aware of their potential challenges and limitations.

- **Bias in algorithms**: AI and ML models learn from the data they are trained on. If the training data contains biases, the models will likely perpetuate and even amplify those biases in their outputs. This can lead to inaccurate or unfair results, especially in sensitive areas like credit scoring or hiring. For example, an AI model trained on historical loan data might inadvertently discriminate against certain demographic groups if the historical data reflects past biases in lending practices.

- **Need for large training datasets**: AI and ML models typically require large amounts of high-quality training data to perform effectively. Obtaining and preparing such datasets can be time-consuming, expensive, and challenging, especially for specialized domains or rare events. For example, training an AI model to detect fraudulent transactions might require a massive dataset of both legitimate and fraudulent transactions, which can be difficult to obtain while protecting sensitive information.

- **Lack of transparency and explainability**: Some AI and ML models, particularly deep learning models, can be complex and opaque, making it difficult to understand how they arrive at their decisions. This lack of transparency can make it challenging to identify and correct errors or biases in the model. For example, if an AI model flags a customer transaction as potentially fraudulent, it might be difficult to understand the specific factors that led to that decision, making it challenging to investigate and resolve the issue.

- **Overfitting to training data**: AI and ML models can sometimes become too closely fitted to the specific characteristics of their training data, leading to poor performance on new or unseen data. This phenomenon, known as overfitting, can limit the generalizability and accuracy of the models. For example, an AI model trained to identify defective products on a specific production line might not perform well when applied to a different production line with slightly different characteristics.

- **Data security and privacy concerns**: Using AI and ML for data quality management often involves processing sensitive data, which raises concerns about data security and privacy. Organizations need to implement robust security measures to protect data from unauthorized access and misuse. For example, AI models used to analyze patient health records must comply with HIPAA regulations to ensure the privacy and security of sensitive health information.

- **Need for ongoing monitoring and maintenance**: AI and ML models are not static; they need to be continuously monitored and maintained to ensure they remain accurate and effective over time. This can involve retraining models with new data, adjusting parameters, and addressing any emerging issues. For example, an AI model used to predict customer churn might need to be retrained periodically with updated customer data to maintain its accuracy as customer behavior changes.

By being mindful of these challenges and taking appropriate measures to mitigate them, organizations can effectively leverage the power of AI and ML to significantly enhance their data quality management efforts.

Challenges in Maintaining Data Quality and Integrity

Maintaining data quality and integrity is an ongoing challenge for organizations, particularly as the volume, complexity, and variety of data continue to grow. Several key challenges make it difficult to ensure high-quality data over time.

Data Volume and Velocity

The sheer volume of data generated by modern systems, combined with the speed at which data is produced (velocity), makes it difficult to maintain data quality at scale. In industries such as e-commerce, telecommunications, and finance, organizations must process and analyze massive amounts of data in real time, making traditional data quality methods insufficient.

To address this challenge, organizations must implement scalable data quality solutions that can handle large datasets and process data in real time. Automation and AI-driven tools are essential for managing data quality at scale.

Complex Data Ecosystems

Many organizations operate with multiple systems, databases, and data sources, both internal and external. Ensuring data quality across these disparate systems is a significant challenge. Data must be synchronized and consistent across different platforms, and integration issues can lead to discrepancies, duplication, or data loss.

To mitigate this challenge, organizations should invest in data integration tools that can automate the process of synchronizing data between systems. Additionally, strong data governance frameworks are necessary to ensure that data is managed consistently across the entire organization.

Evolving Data Standards

Data standards and regulations are constantly evolving, requiring organizations to adapt their data quality practices to meet new requirements. For example, the introduction of GDPR in Europe placed new obligations on organizations to ensure the accuracy and security of personal data. Keeping up with these changes can be resource-intensive and requires ongoing investment in technology and training.

Organizations must stay up-to-date with evolving data standards and regulations, ensuring that their data quality practices remain compliant. This may involve updating validation rules, revising data governance policies, or implementing new technologies to meet regulatory requirements.

Data Silos

In many organizations, data is stored in silos, with different departments or teams maintaining separate datasets that are not easily accessible to others. Data silos can lead to inconsistencies, duplication, and a lack of visibility into the overall quality of the data. For example, sales and marketing teams may have separate databases for customer information, leading to discrepancies in customer records.

To address this challenge, organizations should work to break down data silos by integrating data across departments and ensuring that all teams have access to a single, unified view of the data. Data governance frameworks can help enforce policies that promote data sharing and consistency across the organization.

Resource Constraints

Maintaining data quality requires dedicated resources, including personnel, technology, and budget. Many organizations struggle to allocate sufficient resources to data quality initiatives, particularly as the demand for data-driven insights grows. Without proper investment, data quality efforts may be insufficient to meet the organization's needs.

To overcome this challenge, organizations should prioritize data quality as a strategic initiative, ensuring that the necessary resources are allocated to support ongoing data quality efforts. Automation and AI-driven tools can help reduce the manual effort required to maintain data quality, making it more scalable and cost-effective.

Case Studies on Data Quality and Integrity

Real-world case studies provide valuable insights into the challenges and solutions for maintaining data quality and integrity. By examining how organizations in different industries have addressed data quality issues, you can identify best practices and lessons learned.

Case Study 1: Healthcare Data Quality in a Large Hospital Network

A large hospital network faced significant data quality challenges due to the complexity of its patient records system. Patient data was spread across multiple systems, including electronic health records (EHR), billing systems, and laboratory databases. Inconsistent data entry practices, lack of synchronization between systems, and missing patient information led to operational inefficiencies, incorrect billing, and challenges in patient care coordination.

Solution

The hospital implemented a comprehensive data governance framework that standardized data entry practices across all systems. Data validation rules were enforced at the point of entry, and automated tools were used to identify and correct inconsistencies. Additionally, a master data management (MDM) system was deployed to create a unified view of patient records, ensuring that all departments had access to accurate and complete patient information.

Results

The hospital saw a significant improvement in data quality, with a reduction in errors and missing information. Patient care coordination improved, and billing accuracy increased, resulting in fewer disputes and faster reimbursement from insurance providers.

Case Study 2: E-Commerce Data Quality for Customer Personalization

An e-commerce company struggled with data quality issues related to customer profiles. The company used customer data to personalize marketing campaigns, product recommendations, and promotional offers. However, incomplete and outdated customer information led to ineffective campaigns and missed opportunities for upselling and cross-selling.

Solution

The company implemented automated data cleansing and validation tools to ensure that customer profiles were up-to-date and complete. Machine learning algorithms were used to analyze customer behavior and identify patterns that could be used to predict when customer data was likely to become outdated. The company also launched a data governance initiative to standardize data collection practices and ensure that all customer touchpoints provided consistent data.

Results

The company saw a significant improvement in the accuracy and effectiveness of its marketing campaigns, with higher conversion rates and increased customer engagement. Personalized product recommendations became more relevant, leading to increased sales and customer satisfaction.

Key Stakeholders in Data Quality Management

Maintaining data quality is a collaborative effort that involves multiple stakeholders across an organization. Each stakeholder plays a critical role in ensuring that data remains accurate, consistent, and reliable.

Data Stewards

Data stewards are responsible for overseeing data quality within specific domains or departments. They ensure that data is collected, maintained, and used according to the organization's data governance policies. Data stewards are often the first point of contact for addressing data quality issues and are responsible for resolving discrepancies, ensuring consistency, and promoting best practices.

Data Owners

Data owners are individuals or teams responsible for specific datasets. They have the authority to define data standards, establish validation rules, and make decisions regarding data quality. Data owners work closely with data stewards and IT teams to ensure that data is managed effectively throughout its lifecycle.

IT and Data Management Teams

IT teams and data management professionals play a crucial role in implementing and maintaining the technical infrastructure required to manage data quality. This includes deploying data validation tools, managing databases, and ensuring that data integration processes are robust and secure.

Business Analysts and Decision-Makers

Business analysts and decision-makers rely on high-quality data to generate insights and make informed decisions. They are key stakeholders in ensuring that data quality meets the needs of the business. Analysts often provide feedback on data quality issues and work with data stewards to address any gaps or inconsistencies.

Compliance and Legal Teams

In regulated industries, compliance and legal teams are responsible for ensuring that the organization meets data quality standards required by law. They play a critical role in enforcing data governance policies and ensuring that data quality practices comply with regulatory requirements such as GDPR, HIPAA, and SOX.

By involving all key stakeholders in the data quality management process, organizations can ensure that data quality remains a priority and that efforts are coordinated across all departments and functions.

Emerging Trends in Data Quality and Integrity

As organizations continue to evolve, new trends are emerging that impact how data quality and integrity are managed. These trends are driven by advances in technology, changes in regulatory requirements, and the increasing importance of data in decision-making.

Data Quality as a Service

Data Quality as a Service (DQaaS) is an emerging trend where organizations outsource their data quality management to third-party providers. DQaaS providers offer cloud-based solutions that handle data profiling, cleansing, validation, and monitoring. This approach allows organizations to benefit from advanced data quality tools without the need for significant in-house resources or expertise.

Blockchain for Data Integrity

Blockchain technology is increasingly being explored as a solution for ensuring data integrity. Blockchain provides a decentralized, immutable ledger that can be used to record data transactions, ensuring that data cannot be tampered with once it has been recorded. This makes blockchain particularly useful for industries where data integrity is critical, such as finance, healthcare, and supply chain management.

Real-Time Data Quality Monitoring

As organizations increasingly rely on real-time data for decision-making, there is a growing need for real-time data quality monitoring. This involves continuously validating and cleaning data as it is generated, ensuring that data is accurate and up-to-date at all times. Real-time monitoring is particularly important for industries such as finance and e-commerce, where delays in data processing can result in missed opportunities or financial losses.

AI-Driven Data Quality Solutions

AI and machine learning are playing an increasingly important role in data quality management. AI-driven tools can automatically detect and correct data quality issues, identify patterns and trends, and predict when data is likely to become inaccurate or outdated. As AI technology continues to advance, expect to see even more sophisticated solutions for maintaining data quality at scale.

Privacy-Enhanced Data Quality

With the growing importance of data privacy regulations such as GDPR, organizations are increasingly focusing on privacy-enhanced data quality. This involves ensuring that data quality practices comply with privacy laws and that sensitive data is protected throughout its lifecycle. Techniques such as data anonymization, encryption, and access controls are being integrated into data quality management processes to ensure that data remains both high-quality and secure.

These emerging trends are shaping the future of data quality and integrity, providing organizations with new tools and strategies to manage their data more effectively.

Conclusion

Data quality and integrity are fundamental to the success of any organization that relies on data for decision-making, operations, or compliance. Ensuring that data is accurate, consistent, complete, and reliable requires a comprehensive approach that includes data governance, validation, cleaning, and continuous monitoring.

As data continues to grow in volume, complexity, and importance, organizations must adopt new technologies and best practices to maintain high data quality. Automation, AI, and machine learning are playing an increasingly important role in this effort, providing scalable solutions for managing data quality at scale.

By prioritizing data quality and integrity, organizations can gain a competitive advantage, make more informed decisions, and ensure that their data remains a valuable asset for years to come.

CHAPTER 9

Data Governance and Usability Frameworks

In the modern digital economy, data is often described as one of the most valuable assets a business can possess. This has led to an increased emphasis on both data governance—the frameworks and policies that ensure data is accurate, secure, and used appropriately—and usability—how accessible and easy to use these data systems are for end users. Without effective governance, data becomes fragmented, unreliable, and prone to misuse, which can result in compliance failures and strategic missteps. On the other hand, without an emphasis on usability, even well-governed data can be rendered ineffective if users are unable to access or understand it efficiently.

Data governance and usability are often seen as two distinct disciplines, but they are inherently intertwined. Good governance practices set the foundation for usable data systems, while usability ensures that stakeholders can leverage governed data for decision-making, innovation, and compliance. This expanded chapter explores the critical components of data governance, delves into usability principles as they relate to data systems, and presents a framework for integrating both areas to achieve the highest levels of data quality, security, and usability.

© Saurav Bhattacharya 2025
P. Gujar, *Data Usability in the Enterprise*, https://doi.org/10.1007/979-8-8688-1183-8_9

Importance of Data Governance

As organizations increasingly shift toward data-driven decision-making, the importance of well-structured data governance becomes more pronounced. Data governance ensures that data is managed responsibly, following legal, regulatory, and ethical standards while also promoting accuracy, consistency, and security. Whether in industries such as healthcare, financial services, retail, or technology, data governance provides the scaffolding that supports compliance, data quality, and trust.

The following explains the key benefits of data governance.

- Data quality **and accuracy**: Reliable data governance ensures the accuracy and completeness of data, which is critical for analytics, reporting, and decision-making. Data that is incorrect or incomplete can lead to incorrect decisions, financial losses, and regulatory penalties.

- **Regulatory compliance**: Many industries are governed by strict regulations, such as GDPR, HIPAA, and CCPA. Non-compliance with these regulations can result in significant financial penalties and reputational damage. A robust governance framework ensures that organizations can comply with these regulations by managing data access, use, and protection according to legal requirements.

- **Security and risk management**: Data breaches and unauthorized access to sensitive information can have severe financial and legal repercussions. Data governance frameworks implement policies and technologies to mitigate risks and protect against data breaches and cyber threats.

- **Operational efficiency**: Good governance structures help streamline data management processes, reducing the time and resources required to find, process, and use data. It enables different departments within the organization to work with standardized, high-quality data, increasing efficiency and productivity.

- **Data-driven decision-making**: Data governance enables a culture of data-driven decision-making by ensuring that stakeholders have access to accurate, up-to-date, and well-structured data. This, in turn, allows for better forecasting, reporting, and strategic planning.

As organizations continue to expand their data operations, the need for robust governance frameworks that balance security, compliance, and usability becomes more critical. The following sections outline the key principles that underpin successful data governance.

Principles of Data Governance

Data governance is built on several core principles that ensure the appropriate management of data throughout its lifecycle. These principles include accountability, transparency, data quality, privacy and security, data stewardship, ownership, and risk management.

Accountability

A core tenet of data governance is that individuals or groups within the organization are held accountable for the management and use of data. Data accountability involves clearly defined roles and responsibilities, from **data owners** who are responsible for specific data assets to **data stewards** who ensure that governance policies are enforced on a day-to-day basis.

By establishing accountability at all levels, organizations ensure that data is managed in a way that is aligned with corporate goals and regulatory requirements.

Accountability also extends to the enforcement of data policies, ensuring that stakeholders follow proper procedures for data access, use, and sharing. Clear accountability structures reduce the risk of errors, data breaches, and non-compliance with regulations.

Transparency

Transparency is essential in data governance because it ensures that the processes and decisions around data management are open and understandable to all relevant stakeholders. This includes providing clear documentation on how data is collected, processed, stored, and shared, as well as making governance policies accessible and comprehensible. Transparency fosters trust in the data system, both within the organization and with external stakeholders such as regulators, customers, and partners.

In addition to promoting trust, transparency also aids in identifying and addressing issues such as data inconsistencies, access violations, or non-compliance. It ensures that all actions related to data management can be audited and traced back to accountable parties.

Data Quality

Data quality is one of the most critical aspects of governance. Poor data quality leads to inaccurate reporting, flawed analytics, and misguided decision-making. High-quality data is accurate, complete, consistent, and timely. Governance frameworks should include processes for data validation, cleansing, and monitoring to ensure that data quality remains high across the organization.

To maintain data quality, organizations must establish data quality metrics and conduct regular audits to assess the state of their data. Automating data quality checks through tools that identify and rectify issues in real time is also an effective strategy for maintaining data integrity.

Privacy and Security

In today's regulatory landscape, where data breaches and misuse of personal data can lead to significant financial and reputational damage, privacy and security have become foundational elements of data governance. **Privacy** refers to protecting personal and sensitive information from unauthorized access. **Security** encompasses the tools, policies, and practices used to protect all forms of data from theft, loss, or damage.

Data governance frameworks should include robust security protocols such as encryption, multi-factor authentication, access controls, and data anonymization to ensure that sensitive information is protected throughout its lifecycle. These protocols should also comply with relevant data protection regulations, such as GDPR and HIPAA, to avoid legal liabilities and penalties.

Data Stewardship

Data stewards are individuals responsible for ensuring that the organization's data governance policies are implemented and followed. They play a key role in maintaining data quality, facilitating data access, and ensuring that data usage complies with governance policies. Effective data stewardship ensures that data governance principles are applied consistently across the organization and that all stakeholders adhere to established rules and guidelines.

Data stewards also act as intermediaries between technical teams (e.g., IT departments) and business units, helping to translate governance requirements into actionable processes. They oversee activities like data classification, validation, and reporting, ensuring that governance policies are enforced effectively.

Data Governance Policies

Data governance policies provide the structure and rules for how data should be managed, used, and protected within an organization. These policies define key procedures like the following.

- **Data access controls**: Who can access specific data, and under what conditions?

- **Data classification**: Organizing data into categories based on its sensitivity, value, and regulatory requirements.

- **Data retention**: Determining how long data should be stored and when it should be archived or deleted.

- **Data sharing**: Guidelines for sharing data internally and with third parties, ensuring that privacy and security are maintained.

Governance policies should be regularly reviewed and updated to reflect changes in technology, regulations, and organizational needs. They should also be communicated clearly to all stakeholders to ensure compliance.

Data Ownership

Data ownership is an often overlooked but crucial element of data governance. It involves assigning responsibility for specific data assets to individuals or teams within the organization. Data owners are accountable for ensuring that their data is accurate, secure, and compliant with governance policies. They also have the authority to make decisions about how the data is used, shared, and protected.

Data ownership is particularly important in large organizations with complex data ecosystems, as it ensures that someone is always accountable for the data and that there are clear points of contact for data-related inquiries and issues.

Risk Management in Data Governance

Risk management is a critical aspect of data governance, as organizations must constantly evaluate and mitigate risks associated with data breaches, compliance failures, and operational disruptions. Effective governance frameworks include risk assessments that identify potential vulnerabilities in data management processes, as well as policies for addressing and mitigating these risks.

The following describes risk management strategies in data governance.

- **Regular audits**: Conducting internal and external audits to assess compliance with governance policies and identify areas of improvement.

- **Data breach response plans**: Establishing protocols for responding to data breaches, including notification procedures and recovery strategies.

- **Access controls and monitoring**: Implementing tools that monitor data access and usage to detect suspicious activity and prevent unauthorized access.

By incorporating risk management into data governance, organizations can reduce their exposure to data-related threats and ensure that their data assets are managed responsibly.

Data Governance Frameworks

To implement effective data governance, organizations often rely on established frameworks that provide best practices, policies, and processes for managing data. These frameworks vary in scope and focus, but they all aim to provide a structured approach to ensuring data quality, security, and compliance.

DAMA-DMBoK

The DAMA-DMBoK framework is a widely recognized guide to data management and governance best practices. Developed by the Data Management Association International (DAMA), the Data Management Body of Knowledge (DMBoK) provides a comprehensive set of standards for managing data across various disciplines, including data governance, quality, architecture, and security. The framework is structured around ten core knowledge areas, such as data integration, data warehousing, metadata management, and data governance.

The framework emphasizes the importance of aligning data management activities with business goals and ensuring that data is managed as a valuable asset. It provides organizations with the tools to implement governance policies that support compliance, data quality, and operational efficiency.

ISO/IEC 38500

The ISO/IEC 38500 framework offers governance principles for IT and data management. This international standard provides guidance on how organizations can align their IT and data governance efforts with broader business goals. While the framework covers all aspects of IT governance, it includes specific principles for data governance, such as ensuring accountability, setting clear governance objectives, and promoting responsible use of data.

ISO/IEC 38500 is particularly valuable for organizations seeking to implement a governance framework that supports data-driven innovation while ensuring compliance with regulatory requirements and risk management practices.

COBIT (Control Objectives for Information and Related Technologies)

The COBIT framework, developed by ISACA, is widely used for IT governance and management. COBIT provides a structured approach to aligning IT with business goals, ensuring that data is managed responsibly and in compliance with regulations. The framework includes guidelines for data governance, covering areas such as data lifecycle management, risk assessment, and compliance monitoring.

COBIT also includes performance metrics and maturity models, which help organizations assess the effectiveness of their data governance practices and identify areas for improvement.

NIST Cybersecurity Framework

The NIST Cybersecurity Framework was developed by the National Institute of Standards and Technology (NIST) to provide a risk-based approach to managing cybersecurity threats. While the framework focuses primarily on cybersecurity, it includes important guidelines for data governance, particularly in areas related to protecting sensitive data and ensuring compliance with privacy regulations.

The NIST Cybersecurity Framework is structured around five core functions—identify, protect, detect, respond, and recover—which provide a comprehensive approach to managing data risks and ensuring that data governance policies are aligned with security best practices.

GDPR and Data Governance

The General Data Protection Regulation (GDPR) is one of the most comprehensive and stringent data protection regulations in the world. GDPR mandates that organizations take strong measures to protect personal data, ensuring that it is collected, processed, and stored securely and transparently. Compliance with GDPR requires organizations to implement robust data governance frameworks that protect individuals' rights and ensure that data is handled responsibly.

GDPR has several key provisions related to data governance, including the following.

- **Data subject rights**: Individuals have the right to access, correct, and delete their personal data.

- **Data breach notification**: Organizations must notify regulators and affected individuals of data breaches within 72 hours.

- **Data protection officers (DPOs)**: Organizations that process large volumes of personal data must appoint a DPO to oversee compliance with GDPR.

Data governance frameworks that comply with GDPR must prioritize privacy, security, and transparency while ensuring that data is used responsibly and in accordance with individuals' rights.

CCPA and Data Governance

The California Consumer Privacy Act (CCPA) is a landmark data privacy law in the United States that provides California residents with rights regarding the collection, use, and sale of their personal data. Similar to GDPR, CCPA requires organizations to implement data governance policies that protect personal information and ensure compliance with the law.

Under CCPA, consumers have the right to

- **Know** what personal data is being collected about them.

- **Delete** their personal data, subject to certain exceptions.

- **Opt out** of the sale of their personal data.

Organizations subject to CCPA must develop governance frameworks that provide clear guidelines for data access, transparency, and deletion while protecting consumers' privacy and data rights.

SOX and Financial Data Governance

The Sarbanes-Oxley Act (SOX) is a US federal law that mandates strict standards for financial reporting and data governance in public companies. SOX requires organizations to implement controls that ensure the accuracy and integrity of financial data, as well as policies for auditing and reporting on data management practices.

For organizations in the financial sector, data governance frameworks must include robust audit trails, data retention policies, and access controls to ensure compliance with SOX and protect the integrity of financial data.

HIPAA and Healthcare Data Governance

The Health Insurance Portability and Accountability Act (HIPAA) establishes stringent requirements for protecting personal health information (PHI) in the healthcare industry. HIPAA mandates that healthcare organizations implement governance policies to secure electronic health information, manage data access, and ensure privacy.

Data governance in healthcare must prioritize the following.

- **Patient privacy**: Ensuring that PHI is protected from unauthorized access.

- **Data security**: Implementing strong security measures, such as encryption and access controls, to protect health data.

- **Compliance**: Ensuring that data management practices comply with HIPAA regulations.

Healthcare organizations must also develop governance policies that support data sharing between providers while maintaining compliance with HIPAA.

FAIR Data Principles

The FAIR Data Principles—Findable, Accessible, Interoperable, and Reusable—are a set of guidelines designed to improve the management and accessibility of data in research and academia. These principles promote open science and data sharing by ensuring that data is properly structured, documented, and available for reuse by others.

The FAIR principles are increasingly being adopted by organizations outside of academia because they provide a framework for improving data transparency and usability. Organizations that implement FAIR data governance frameworks can improve the accessibility and interoperability of their data, making it easier to share and collaborate on data-driven projects.

Data Capability Assessment Model

The Data Capability Assessment Model (DCAM), developed by the EDM Council, provides a comprehensive framework for assessing and improving data management capabilities within an organization. DCAM focuses on evaluating key areas such as data governance, architecture, quality, and analytics to help organizations develop effective data strategies.

DCAM is particularly useful for organizations looking to improve their data governance practices by identifying gaps in their current data management processes and implementing best practices to address these gaps.

Usability in Data Governance

While data governance ensures the accuracy, security, and compliance of data, usability focuses on making data systems intuitive, accessible, and user-friendly. In the context of data governance, usability is critical because it ensures that stakeholders can efficiently access, interpret, and act upon governed data. Without good usability, even well-governed data may remain underutilized or difficult to access.

Usability in data governance involves designing systems and processes that prioritize user experience, ease of use, and efficiency. This includes everything from the design of data access tools and dashboards to the organization of data repositories and the implementation of search and retrieval functions.

Defining Usability in the Context of Data Governance

In the context of data governance, **usability** refers to how easily users— whether they are data scientists, business analysts, or executives—can access, navigate, and interact with the data systems in place. Usability is concerned with the design of data access points, interfaces, and tools that facilitate user interaction with data, ensuring that users can perform tasks quickly, efficiently, and with minimal errors.

Usability also extends to how intuitive and user-friendly data systems are for non-technical stakeholders. For example, business users should be able to access and use data without needing to rely on IT teams for assistance. Usable data systems empower users across the organization to make data-driven decisions by providing them with the tools and interfaces they need to interact with data effectively.

Usability Principles in Data Systems

Several core principles of usability should be applied to data governance systems.

- **Learnability**: Data systems should be easy for new users to learn, with intuitive interfaces and clear instructions. This reduces the time required for onboarding and helps users become productive more quickly.

- **Efficiency of use**: Users should be able to perform common tasks—such as data retrieval, analysis, and reporting—efficiently without encountering unnecessary obstacles or delays.

- **Memorability**: Once users have learned how to interact with the system, they should be able to return to it after a period of non-use without having to relearn how to navigate or use it.

- **Error prevention**: Data systems should be designed to prevent user errors, particularly in areas like data modification or deletion. Error prevention features include confirmation dialogs, input validation, and clear warnings for potentially irreversible actions.

- **Satisfaction**: Users should find the system satisfying and enjoyable to use. A positive user experience reduces frustration and increases system adoption, ensuring that data systems are used consistently and effectively across the organization.

These principles are critical to the success of any data governance system, as they ensure that stakeholders can interact with data systems in a way that is both efficient and satisfying.

Usability Evaluation Methods

Evaluating the usability of data governance systems is essential to ensuring that they meet the needs of users. Several evaluation methods can be used to assess usability, including the following.

- **User testing**: User testing involves observing real users as they interact with the data system, identifying pain points, inefficiencies, and areas for improvement. This method provides valuable insights into how users experience the system and what can be done to improve usability.

- **Heuristic evaluation**: In a heuristic evaluation, usability experts review the system against established usability heuristics (such as **Nielsen's 10 Usability Heuristics**) to identify potential usability issues. This method can be used early in the design process to catch usability problems before the system is deployed.

- **Surveys and feedback forms**: Gathering feedback from users through surveys and feedback forms provides insights into their experiences with the system, allowing organizations to identify areas for improvement.

- **Task efficiency metrics**: Measuring how long it takes users to complete specific tasks—such as retrieving data, generating reports, or modifying records— provides quantitative data on the system's efficiency. Long task times may indicate usability issues that need to be addressed.

By using these evaluation methods, organizations can ensure that their data governance systems are optimized for usability and meet the needs of all stakeholders.

Information Architecture and Usability

The information architecture of a data system plays a critical role in its usability. **Information architecture** refers to how data is organized, labeled, and structured within a system. It has a direct impact on how easily users can find, access, and navigate through the data. Poor information architecture can lead to data silos, confusion, and inefficiencies, making it difficult for users to work with governed data.

Effective information architecture involves the following.

- **Data hierarchy**: Organizing data into logical categories, subcategories, and hierarchies that reflect how users expect to find and use information.

- **Labeling and metadata**: Ensuring that data fields, categories, and records are clearly labeled and accompanied by metadata that provides context and explanations. Metadata makes data easier to search, navigate, and understand.

- **Search and navigation**: Providing robust search functionality that allows users to quickly locate relevant data, even in large datasets. Effective search tools and navigation menus improve the overall user experience and make data systems more usable.

Human-Computer Interaction in Data Governance

Human-computer interaction (HCI) focuses on how humans interact with computers and data systems, with an emphasis on designing interfaces that are intuitive, efficient, and user-friendly. In the context of data governance, HCI plays a crucial role in ensuring that users can interact with data in a way that is both effective and satisfying.

The following are key HCI considerations for data governance systems.

- **User-centered design**: Designing data systems around the needs and behaviors of the users rather than forcing users to adapt to complex or unintuitive interfaces.

- **Interactivity**: Providing interactive tools and features—such as drag-and-drop dashboards, customizable reports, and real-time data visualization—that allow users to manipulate and analyze data in ways that are meaningful to them.

- **Feedback**: Providing clear, immediate feedback to users about their actions within the system. For example, if a user uploads new data or generates a report, the system should provide feedback confirming that the action was successful or alert the user to any errors.

By incorporating HCI principles into data governance systems, organizations can improve usability and ensure that users can interact with data seamlessly and efficiently.

Data Democratization and Usability

Data democratization is the process of making data accessible to all employees, regardless of their technical expertise. The goal of data democratization is to enable a broader range of users—beyond data scientists and IT professionals—to access and use data for decision-making, problem-solving, and innovation.

Usability plays a key role in data democratization by ensuring that data systems are designed to be intuitive and user-friendly for non-technical users. This involves creating easy-to-use interfaces, providing training and support, and implementing role-based access controls to ensure that users can access the data they need without being overwhelmed by unnecessary complexity.

Building a Data Governance Usability Framework

To maximize the effectiveness of data governance, organizations must integrate usability principles into their governance frameworks. A **data governance usability framework** combines the best practices of data governance with user-centered design, ensuring that stakeholders can easily access, interpret, and act upon governed data.

Integration of Governance and Usability

Integrating governance and usability requires collaboration between various stakeholders, including data stewards, IT teams, business users, and user experience designers. By incorporating usability considerations into governance policies, organizations can create data systems that are both secure and user-friendly.

Structuring a Usability-Oriented Governance Framework

A usability-oriented governance framework includes several key components.

- **User-centered policies**: Policies should account for how different users interact with data. For example, data scientists may need access to raw data, while business users may require aggregated data presented in dashboards or reports.

- **Role-based access controls (RBAC)**: Implementing RBAC ensures that users have access to the data they need while maintaining security and compliance.

- **Training and support**: Providing ongoing training and support for users ensures that they can effectively use data systems. This includes offering user guides, tutorials, and help desk support.

- **Iterative design**: Data systems should be designed and refined iteratively based on user feedback and changing business needs. Regular updates ensure that the system remains user-friendly and aligned with governance goals.

Key Metrics for Evaluating Usability

To ensure that a governance framework is meeting usability goals, organizations should track key metrics such as the following.

- **Time to data access**: Measuring how long it takes users to find and access the data they need.

- **Error rates**: Tracking the frequency of user errors, such as incorrect data entry or failed searches.

- **User satisfaction scores**: Gathering feedback from users to assess their overall satisfaction with the data system.

- **Adoption rates**: Monitoring how frequently different user groups access and use the system.

These metrics provide valuable insights into how well the data governance framework is serving the needs of users and where improvements may be needed.

Balancing Compliance and Usability

Finding the right balance between compliance and usability can be challenging, as strict governance policies can sometimes hinder user experience. To achieve the right balance, organizations should do the following.

- **Assess risk levels.** Identify which datasets require the highest levels of security and which can prioritize usability without compromising compliance.

- **Simplify compliance processes.** Streamline workflows for compliance-related tasks, such as data access requests and audit reporting.

- **Automate governance checks.** Use automated tools to monitor compliance with governance policies, reducing the burden on users.

Customizing Governance for Different Stakeholders

Different stakeholders within an organization have different needs when it comes to accessing and using data. A governance framework should be customized to accommodate these needs, providing tailored access, tools, and interfaces for different user groups.

The following are some examples.

- **Executives** may need access to high-level reports and dashboards that provide insights into business performance.

- **Data analysts** may require access to detailed, raw data for analysis and reporting.

- **Legal teams** may need access to audit trails and compliance documentation.

By tailoring the governance framework to meet the needs of different stakeholders, organizations can improve usability while maintaining strong governance controls.

Governance for Structured vs. Unstructured Data

Data governance frameworks must account for the differences between structured and unstructured data. **Structured data** is highly organized and typically stored in databases. **Unstructured data**—such as emails, social media posts, and documents—does not follow a predefined format.

Governance frameworks for structured data often focus on ensuring data accuracy, consistency, and compliance with regulatory requirements. In contrast, governance for unstructured data involves managing data classification, storage, and retrieval in a way that makes unstructured data accessible and usable for analysis.

Data Lifecycle Management and Usability

Data lifecycle management refers to the process of managing data from its creation and use to its eventual archiving or disposal. Usability plays a key role in ensuring that users can access and use data effectively throughout its lifecycle.

The following are some key stages of the data lifecycle.

- **Data creation**: Ensuring that data is captured accurately and consistently.

- **Data storage**: Organizing data in a way that makes it easy to retrieve and use.

- **Data usage**: Providing users with the tools and interfaces they need to analyze and act on the data.

- **Data archiving**: Ensuring that archived data is still accessible for future use.

- **Data disposal**: Implementing policies for securely disposing of data that is no longer needed.

By designing data systems that support usability at each stage of the data lifecycle, organizations can ensure that data remains accessible and useful throughout its lifespan.

Usability in Data Retention and Disposal

Governance policies often include rules for how long data should be retained before it is archived or deleted. Usability considerations come into play when designing systems for managing data retention and disposal, ensuring that users can easily identify which data should be retained and which should be disposed of.

For example, data retention systems should provide clear labels and metadata to help users understand the status of different datasets and when they are due for deletion or archiving. Automated tools can also help enforce data retention policies, reducing the risk of human error.

Case Studies in Data Governance and Usability

Financial Services

In the financial services industry, data governance is essential for complying with regulations like Sarbanes-Oxley (SOX), Dodd-Frank, and Basel III. Financial institutions must maintain rigorous control over their data to prevent fraud, ensure transparency, and protect sensitive customer information.

The following are some usability considerations.

- Financial analysts rely on fast, accurate access to data for decision-making and regulatory reporting. If the data systems are difficult to navigate or access, it can delay critical business processes.

- A well-governed but unusable system may lead to data silos, as employees find workarounds to access data outside of approved channels, creating shadow IT systems.

Healthcare

Healthcare organizations handle sensitive patient data, which must comply with regulations like HIPAA. These regulations place strict requirements on how data is stored, accessed, and shared.

The following are usability considerations.

- Doctors and healthcare providers need to access patient data quickly and accurately, often in time-sensitive situations. If the systems are cumbersome or slow, it can have real-life consequences.

- Usable data governance systems ensure that providers can easily access patient records while maintaining the highest levels of data privacy and security.

Retail and E-commerce

In the retail and e-commerce sectors, data governance is essential for managing vast amounts of customer data, including purchase histories, preferences, and payment information. Compliance with privacy regulations like GDPR and CCPA is crucial for protecting customer data.

The following are usability considerations.

- Customer service representatives need quick access to order histories, while marketers require insights into consumer behavior. Usability ensures that both roles can access the data they need without violating governance policies.

- Usability in data governance also helps prevent **data fatigue**, where employees are overwhelmed by the sheer volume of data available.

Government and Public Sector

Government organizations must adhere to strict data governance policies related to transparency, accountability, and citizen privacy. Regulations like the Freedom of Information Act (FOIA) and General Data Protection Regulation (GDPR) require governments to store and manage data in secure, compliant ways.

The following are usability considerations.

- Government employees need easy access to public records but must also ensure that citizens' privacy is protected. Usable data systems ensure that employees can retrieve information efficiently, while data governance policies ensure compliance with laws.

- Transparency is critical in the public sector. Governance frameworks must ensure that data is accessible to the public while preventing unauthorized access to sensitive information.

Technology Companies

Technology companies often handle vast amounts of user data, requiring strict governance policies to comply with privacy regulations and prevent security breaches. Companies like Google, Facebook, and Amazon manage enormous data stores that contain personal, financial, and behavioral data from billions of users.

The following are usability considerations.

- Engineers, data scientists, and product managers need to access large datasets for product development and analytics. Usability ensures that these employees can efficiently access and analyze data, while governance ensures that user privacy and security are maintained.

- Balancing the need for innovation and development with the requirement for privacy compliance is a challenge that technology companies face daily.

Energy Sector

In the energy sector, data governance plays a critical role in managing operational data related to power generation, transmission, and consumption. Energy companies collect vast amounts of data from sensors, smart meters, and other sources, making it essential to implement strong governance policies to ensure data accuracy and reliability.

The following are usability considerations.

- Engineers and technicians need to quickly access real-time data from energy grids to make critical decisions about power distribution and maintenance.

- Usability in data governance ensures that this data is presented in a way that is easy to understand and act upon, particularly in time-sensitive situations.

Higher Education

Universities and research institutions generate and manage large volumes of data related to students, research projects, and administration. Data governance in higher education must prioritize compliance with privacy regulations such as FERPA (Family Educational Rights and Privacy Act) while ensuring that data is accessible for academic research and decision-making.

The following are usability considerations.

- Researchers need access to large datasets for analysis, while administrative staff require tools for managing student records and financial data.

- Usable governance systems ensure that these different groups can access the data they need without compromising security or privacy.

Telecommunications

Telecommunications companies collect vast amounts of customer data, including call records, Internet usage, and location data. Data governance in the telecom industry is essential for ensuring that this data is used responsibly and complies with privacy regulations such as CCPA and GDPR.

The following are usability considerations.

- Customer service teams need access to data related to service usage, billing, and account history to assist customers effectively.

- Usable data governance systems ensure that customer service representatives can access relevant data quickly and efficiently while maintaining compliance with privacy regulations.

The Role of Emerging Technologies in Data Governance and Usability

Emerging technologies are reshaping how organizations handle data governance and usability. From AI and blockchain to IoT and quantum computing, these technologies offer new opportunities and challenges for managing and securing data.

Artificial Intelligence and Machine Learning

Artificial intelligence (AI) and machine learning are increasingly being integrated into data governance frameworks to automate processes, enhance data quality, and improve decision-making. For instance, AI can automate data classification, identifying sensitive data and ensuring it is governed according to company policies.

AI-driven insights and predictions must be delivered to users in a way that is easy to understand and actionable. Integrating AI with usable dashboards and reporting tools ensures that business users can extract meaningful insights from the data without needing deep technical expertise.

Blockchain

Blockchain technology provides a decentralized, transparent way of storing data, making it easier to track data provenance and verify data integrity. Blockchain's distributed ledger technology offers a way to ensure that data is immutable and tamper-proof.

While blockchain offers enhanced security and transparency, it can be complex to implement and use. Ensuring that blockchain-based data systems are intuitive and user-friendly is crucial for encouraging adoption.

Cloud Computing

Cloud computing has revolutionized how organizations store and manage data. By moving to the cloud, companies can scale their data infrastructure rapidly and cost-effectively while ensuring that data remains accessible from anywhere.

Cloud data governance frameworks must ensure that users can easily access data while maintaining control over data security, privacy, and compliance. Intuitive cloud interfaces, combined with robust role-based access controls, ensure that users can efficiently work with data without compromising governance policies.

Data Lakes and Usability

Data lakes offer a flexible, scalable way to store large amounts of structured and unstructured data. However, the sheer size and complexity of data lakes can pose usability challenges, as users may struggle to find and retrieve the data they need.

Effective data governance frameworks for data lakes should include strong metadata management and search capabilities that allow users to locate relevant data quickly. Intuitive data catalogs, coupled with governance policies, ensure that users can navigate the data lake efficiently.

The Internet of Things and Data Governance

The Internet of Things (IoT) generates massive amounts of data from connected devices, raising new challenges for data governance. Ensuring the security and privacy of IoT data is critical, particularly as devices become more ubiquitous in homes, workplaces, and cities.

IoT governance frameworks must balance usability and security. Users, such as system administrators, need clear and intuitive interfaces to manage IoT devices and access the generated data while maintaining robust security measures to prevent unauthorized access.

Quantum Computing and Future Data Security

Quantum computing promises to revolutionize computing power, but it also presents new risks to data security. Quantum computers could potentially break existing encryption methods, making current data security practices obsolete.

Quantum computing's impact on data governance and usability remains to be seen. However, as quantum technologies evolve, organizations must consider how to make quantum-secure data governance systems usable and accessible to stakeholders.

Edge Computing and Data Governance

Edge computing involves processing data closer to where it is generated, such as on IoT devices or local servers, rather than in centralized cloud data centers. Edge computing reduces latency and bandwidth usage, making it ideal for applications that require real-time data processing, such as autonomous vehicles and industrial automation.

Edge computing introduces new challenges for data governance, as data is processed and stored across a decentralized network of devices. Usability is critical in ensuring that data governance policies can be implemented and enforced consistently across these distributed systems, allowing users to access and manage data efficiently.

Robotic Process Automation in Data Governance

Robotic Process Automation (RPA) involves using software robots to automate repetitive tasks, such as data entry, reporting, and compliance monitoring. RPA can enhance data governance by automating data classification, validation, and audit processes, ensuring that data governance policies are followed consistently.

RPA tools must be designed with usability in mind to ensure that users can easily configure, monitor, and manage automated workflows. User-friendly RPA interfaces reduce the complexity of automation, allowing non-technical users to leverage automation tools effectively.

Future Trends in Data Governance and Usability

As the data landscape continues to evolve, new trends are emerging that will shape the future of data governance and usability. These trends are driven by the increasing volume of data, the growing complexity of compliance requirements, and the ongoing demand for innovation in data management technologies.

Data-Centric Business Models

More organizations are shifting to data-centric business models, where data is treated as a strategic asset. This shift requires a governance framework that not only protects data but also enhances its usability, enabling organizations to monetize and leverage data in new ways.

Organizations will need to integrate usability into their data governance strategies to ensure that all employees, from data analysts to executives, can efficiently access and use data for decision-making and innovation.

Zero-Trust Architectures and Data Governance

Zero-trust architectures (ZTAs) are gaining traction as a security model where no user or device is trusted by default, even within the network perimeter. ZTAs enforce strict identity verification and access controls for every user and device attempting to access data.

Data governance frameworks will need to evolve to support zero-trust principles, ensuring that access to sensitive data is controlled and monitored in real time without compromising usability.

Ethical Data Governance

As data collection becomes more pervasive, organizations must consider the ethical implications of how they collect, store, and use data. Ethical data governance requires transparency, accountability, and respect for user privacy rights, particularly in areas like AI and big data analytics.

Governance frameworks must not only comply with legal requirements but also incorporate ethical guidelines to ensure that data is used responsibly. Usability considerations will need to ensure that users are informed about how their data is used and have the tools to manage their privacy preferences.

Data Sovereignty and Cross-Border Data Transfers

With the rise of global data protection regulations, data sovereignty—the principle that data is subject to the laws of the country where it is collected—has become a growing concern. Organizations that operate across multiple jurisdictions must navigate complex rules around cross-border data transfers.

Governance frameworks must account for data localization laws and provide clear processes for ensuring compliance with regional regulations. Usability considerations ensure that legal and compliance teams can easily track where data is stored and how it is being shared across borders.

Automation in Data Governance

As organizations handle increasingly large datasets, manual data governance processes become unsustainable. Automation tools powered by AI and machine learning are emerging to automate tasks like data classification, risk assessment, and compliance monitoring.

Automated governance tools improve both the efficiency and effectiveness of governance frameworks. Usability is crucial in ensuring that these tools are accessible to non-technical stakeholders and that the insights they generate are easy to interpret and act upon.

Data Mesh and Decentralized Data Ownership

The data mesh is an emerging architectural paradigm that decentralizes data ownership by assigning responsibility for data management to individual business domains. Instead of a centralized data lake, each domain manages its own data as a product, with governance policies embedded into the data's lifecycle.

Data governance frameworks will need to adapt to support decentralized data ownership models, ensuring that governance policies are consistently applied across domains while maintaining usability for stakeholders.

Self-Service Analytics and Usability

Self-service analytics tools empower business users to access, analyze, and visualize data without needing assistance from IT or data teams. As more organizations adopt self-service analytics, governance frameworks must ensure that users have access to reliable, governed data while maintaining usability.

Usability will be a key factor in the success of self-service analytics, as users will need intuitive interfaces, real-time data access, and flexible tools to conduct their analyses without compromising governance policies.

ESG Data Governance

Environmental, social, and governance (ESG) data governance is becoming increasingly important as organizations are held accountable for their environmental and social impacts. ESG data governance involves managing data related to sustainability, corporate responsibility, and ethical business practices.

As organizations are required to report on ESG metrics, data governance frameworks must be expanded to include ESG data, ensuring that this data is accurate, transparent, and usable for both internal decision-making and external reporting.

Conclusion

The integration of data governance and usability frameworks is essential for organizations that want to leverage data as a strategic asset while ensuring compliance with regulatory requirements. Data governance ensures that data is accurate, secure, and used responsibly, while usability ensures that stakeholders can efficiently access, interpret, and act upon that data.

In an era where data is both a key driver of business success and a source of potential risk, organizations must invest in governance frameworks that prioritize both security and user experience. By incorporating usability into their data governance strategies, organizations can maximize the value of their data, improve decision-making, and ensure compliance with evolving regulations.

As new technologies and trends continue to shape the data landscape, organizations must remain agile and forward-thinking in their approach to data governance and usability. By staying ahead of emerging trends and adapting their governance frameworks to meet the demands of a data-driven world, businesses can ensure that their data remains a powerful asset for years to come.

CHAPTER 10

Data Privacy and Security Concerns in Software and Data Usability

The digital world's rapid expansion presents a profound duality. Data, the lifeblood of this digital age, holds the keys to unlocking advancements that could revolutionize industries, transform lives, and help us solve our most pressing societal challenges. This treasure trove of information allows us to tailor products and services, personalize experiences, and streamline processes, ultimately boosting efficiency and driving innovation. The potential for positive impact is undeniable. Yet, this progress comes with a paradox: the very data that empowers us also presents unprecedented risks to privacy, security, and autonomy.

Yet, this unprecedented access to information also casts a long shadow, raising concerns about unwarranted surveillance, insidious data misuse, and the erosion of fundamental privacy rights. The same data that empowers us can also be weaponized, used to manipulate, discriminate, or exploit. With every click, every purchase, every social media interaction, we leave a digital trail that can be harvested, analyzed,

P. Gujar, *Data Usability in the Enterprise*, https://doi.org/10.1007/979-8-8688-1183-8_10

and potentially used against us. The risks are real, and the consequences can be devastating.

The quest to achieve a harmonious balance between data privacy, security, and usability is not a mere technical hurdle; it's an ethical imperative that defines the trajectory of our increasingly interconnected future. Striking this delicate balance requires a multifaceted approach that encompasses robust regulations, transparent data practices, and user-centric design principles. It calls for a collective commitment from governments, corporations, and individuals alike to safeguard sensitive information, uphold ethical standards, and empower users to control their digital footprints.

This is a pivotal moment in human history. The choices we make today will determine whether the digital age becomes a force for good, a tool for oppression, or something in between. By prioritizing data privacy, security, and usability, we can harness the power of information to drive progress while protecting the rights and freedoms that are essential to a just and equitable society. The path forward is fraught with challenges, but the stakes are too high to ignore.

Understanding Data Privacy and Security
Definitions and Key Concepts

Let's think beyond data privacy as simply a set of rules and instead focus on it as an extension of individual liberty. The lines between our online and offline selves continue to blur. Data privacy allows you to maintain some control over this digital reflection to decide what aspects remain obscured and which ones we willingly share. When that right is eroded, so too is our sense of agency. This is why safeguarding our personal information isn't just about preventing potential misuse; it's about preserving the freedom to curate our digital identity and ensuring that our virtual persona aligns with our true selves.

Data security is the shield that protects this freedom of choice. It's about robust encryption, yes, but also proactive threat detection, strict incident response protocols, and even employee education about the everyday behaviors that can put sensitive data at risk. Security measures must be woven into the fabric of an organization to ensure privacy isn't an afterthought but a prerequisite for engaging in the digital space. This holistic approach to security ensures that guarding user data isn't simply a compliance checkbox but is embedded into every aspect of a company's operations, fostering a culture where privacy is paramount.

Legal Framework and Global Regulations

The legal realm is where our collective desire for security begins to take shape. However, it's vital to acknowledge that the regulatory landscape still resembles a patchwork quilt. Different countries, even different regions within a single country, often have their own sets of rules, making compliance challenging for multinational businesses. This patchwork nature of data privacy laws means that companies often have to navigate a labyrinthine system, where understanding the nuances of different legal requirements becomes as important as the actual implementation of protective measures.

While these rules often seem to lag behind the pace of tech innovation, they are not without teeth. Consider the substantial fines imposed under the European Union's General Data Protection Regulation (GDPR) for breaches or mishandling of user data. These actions signal a shift; the cost of failing to prioritize privacy and security now has the potential to be far more than mere reputational harm. Such financial penalties serve as a clear deterrent, incentivizing organizations to invest in robust data security measures to avoid both monetary losses and the potential long-term damage to their brand and customer trust.

Impact of Breaches

It's tempting to think that data breaches "just happen" and are an unavoidable evil. This mindset is dangerous. Every publicized breach sparks renewed public scrutiny and often leads to calls for tighter legislative action. The ripple effects of a single breach can extend far beyond the immediate victims, as they fuel a broader debate about the adequacy of existing safeguards and the responsibility of companies to shield user data.

While a single breach may be contained, repeated failures across industries make users wary and hesitant to fully embrace digital services that could otherwise greatly improve their lives. The cumulative impact of multiple breaches erodes public confidence in the digital ecosystem, leading to skepticism and reluctance to share personal information online. If organizations remain complacent, the erosion of trust could lead to an environment where overregulation stifles innovation rather than fosters it. A climate of distrust could prompt governments to impose overly stringent regulations that, while aimed at protecting consumers, could inadvertently hinder technological progress and limit the potential benefits of the digital age.

Data Usability: Importance and Challenges
Definition and Significance

Usability is where the true potential of data lies. Imagine healthcare breakthroughs driven not by small, isolated studies but by analyzing vast datasets to identify patterns and correlations that individual researchers simply couldn't perceive. Think of educational platforms that tailor learning experiences in real-time, meeting students where they are instead of relying on a one-size-fits-none approach.

Usability goes beyond efficiency; it's about making data actionable and accessible. When data is easy to understand and apply, it can drive innovation, inform decision-making, and even change lives. Whether it's a business gaining insights into customer behavior or a researcher uncovering hidden trends, the ability to leverage data effectively hinges on how usable it is. Usability empowers organizations to deliver experiences that feel valuable, intuitive, and even delightful, fostering higher engagement and loyalty.

Balancing Usability with Privacy/Security: The Core Challenge

The inherent tension lies in the fact that greater usability often hinges on collecting and processing more data. This creates an internal conflict within many organizations: the desire to personalize and innovate runs up against the ethical and legal responsibility to minimize data collection and protect the information they gather.

The pursuit of personalized experiences, targeted recommendations, and predictive analytics can tempt organizations to amass large amounts of personal data. However, this very act can trigger privacy concerns, erode trust, and even result in legal repercussions. Striking a balance between these competing interests is essential for any organization that wants to harness the power of data while maintaining ethical practices. This challenge isn't insurmountable. The key is to adopt a 'privacy by design' mentality. Usability cannot be tacked on after the fact; respecting user privacy needs to be ingrained in the system architecture from its earliest conceptualization.

Technical Measures for Balancing Privacy, Security, and Usability

Encryption and access control mechanisms are table stakes, but true progress lies in newer territories. Differential privacy methods allow insights to be gleaned from datasets without revealing individual records. Federated learning permits models to be trained on data without that data ever leaving users' devices, upending traditional notions of centralized data pools.

The traditional safeguards of encryption and access control are essential, but they're just the beginning. The future of data privacy and usability lies in cutting-edge technologies that enable data utilization while protecting individual privacy. Differential privacy and federated learning offer innovative ways to extract valuable insights from data without compromising the confidentiality of personal information. These emerging approaches are reshaping the landscape of data analysis and opening new avenues for responsible innovation.

User-Centric Design

Companies often make the mistake of assuming data privacy and usability are at odds. In fact, well-designed privacy interfaces can enhance both. Simple dashboards allow granular control over what data is shared and for what purposes, building trust. Pairing this with clear explanations of how that data benefits the user creates a true value exchange.

Instead of viewing privacy as a barrier to usability, companies should embrace it as an opportunity for transparent and empowering design. User-friendly privacy interfaces, such as customizable dashboards and clear explanations of data usage, not only protect user data but also foster a sense of trust and collaboration. By involving users in the decision-making process and demonstrating the value they receive in exchange for their data, companies can cultivate a positive relationship with their audience.

The Power of AI

When used responsibly, AI systems can be a potent line of defense. Machine learning excels at spotting anomalies that human teams might miss—think unusual login patterns or attempts to access data from unexpected locations. But let's not forget, these systems are only as good as the data they're trained on, and we must be vigilant against introducing bias that could lead to legitimate users being flagged or false positives that distract security teams.

The analytical prowess of AI can be harnessed to bolster security measures, identifying potential threats with greater accuracy and efficiency than traditional methods. However, the ethical implications of AI in data privacy and security cannot be overlooked. It's crucial to address potential biases in AI algorithms and ensure that they are trained on diverse and representative datasets to avoid discriminatory outcomes and maintain the integrity of security systems.

Ethical Considerations and Future Directions

The Rise of the Privacy-Conscious User

The digital landscape is undergoing a profound shift as societal expectations surrounding data evolve. People are becoming increasingly aware of the potential benefits and harms associated with their personal information. As a result, there's a growing demand for products and services that prioritize privacy. This isn't just a niche concern for the tech-savvy; it's a defining factor in market dynamics and redefines consumer choices. Businesses that fail to adapt to this new reality risk losing consumer trust and market share.

Beyond Compliance, Toward Transparency

While regulations play a crucial role in establishing a baseline for data protection, true trust is built on going above and beyond the bare minimum. Businesses that are transparent about their data practices and actively seek user feedback gain a competitive advantage. This means providing clear explanations about what data benefits the user versus solely the organization. It also means respecting the right to be forgotten, allowing individuals to fully revoke their data if they choose. This approach not only fosters trust but also empowers users to make informed decisions about their digital footprint.

Imagining a More Harmonious Future

This is not a dystopian vision of the future. Instead, let's envision a world where privacy-preserving technologies are ubiquitous. A world where users have granular, easily understandable control over their personal data, and businesses can still provide exceptional experiences without compromising on security, where regulators provide clarity while encouraging innovation, and where society as a whole agrees on fundamental ethical standards regarding the collection and use of data. This harmonious equilibrium is within reach, but it requires a concerted effort from all stakeholders.

Cross-Industry Collaboration

The challenges of the digital age cannot be solved in isolation. Security experts, privacy advocates, software engineers, and even social scientists need to collaborate to build frameworks that balance the needs of users and businesses from the outset. This interdisciplinary approach can lead to the development of innovative solutions that preserve privacy without stifling innovation.

Education is Essential

To navigate the complexities of the digital world, users need to be empowered with the knowledge to make informed decisions. This goes beyond technical literacy and includes media literacy, which helps people understand how data can be used for both good and ill purposes. By equipping individuals with the tools to critically evaluate information and understand the implications of their online activities, we can foster a more privacy-conscious society.

Global Standards

While regional differences will always exist, striving for a harmonized understanding of basic privacy rights is essential. This simplifies things for businesses operating internationally and gives users peace of mind that their data isn't subject to drastic changes in protection when they cross virtual borders. By working toward global standards, we can create a more consistent and predictable privacy landscape that benefits both individuals and organizations.

Case Study: Healthcare

The coming of the digital revolution in the healthcare domain will define the way how health services are administered in the near future. The change is not in the form of adopting new technologies but rather a change at the root level of healthcare; it is in the approach toward patient care, data management, and service delivery. The critical enablers for this evolution are the creation and use of very large volumes of data. This has changed since electronic health records (EHRs), telemedicine, and mHealth applications have become the new normal, with huge volumes of sensitive patient data drawn through healthcare systems. It does come with endless benefits, from better health outcomes and improved management of patients to streamlined operations and personalized care. This transformation promises to revolutionize healthcare delivery, making it more efficient, accessible, and patient-centric.

A digital revolution in healthcare also brings significant challenges. The proliferation of digital tools and the vast amounts of sensitive patient data generated raise concerns about privacy and security. EHRs, telemedicine platforms, and mHealth applications, while beneficial, also pose risks of data breaches and unauthorized access. Safeguarding this sensitive information becomes paramount to maintain patient trust and ensure the integrity of the healthcare system. This necessitates robust cybersecurity measures and stringent adherence to data protection regulations.

However, a digital revolution has given rise to huge challenges, especially in the effort to maintain the secrecy of such information from breaches and unauthorized access. Healthcare is processing so much private data, so this area brings a lot of money to those who target it to bring out cyber threats. The challenge, therefore, is to adopt the new technology while at the same time ensuring it is secure, reliable, and, hence, the best fit for the patient. The digital landscape of healthcare is a lucrative target for cybercriminals due to the wealth of personal and financial information it holds. Striking a balance between technological innovation and robust security measures is crucial to mitigate the risks associated with data breaches and unauthorized access. Ensuring the confidentiality, integrity, and availability of patient data is essential to shield patient privacy and maintain the credibility of healthcare institutions.

The change raises concerns about shared information because patients share the most personal information related to their health, assuming it to be within the tightest confidence and security. Any breach can have vast consequences for the individuals affected and the trust within the healthcare system. Patients entrust healthcare providers with their most intimate health details, expecting the utmost confidentiality and security. A breach of this trust can have devastating consequences, not only for the individuals whose information is compromised but also for the reputation and credibility of the entire healthcare system. The potential fallout from a data breach underscores the importance of implementing comprehensive security protocols and educating both healthcare providers and patients about the risks and responsibilities associated with digital health data.

Healthcare providers operate in an increasingly regulated and standardized environment when it comes to this fast-evolving digital ecosystem. This is mostly in compliance with laws such as HIPAA in the United States, GDPR in Europe, and other regional regulations that dictate the rules according to which patient data must be handled to set a bar for privacy and security in digital healthcare. These regulations impact daily operations for healthcare providers, requiring them to implement strict security protocols, obtain patient consent for data use, and ensure the confidentiality of health information across all digital platforms.

The rapid evolution of digital technologies in healthcare necessitates a robust regulatory framework to ensure the protection of patient data and maintain ethical standards. Laws like HIPAA and GDPR are crucial in establishing guidelines for the collection, storage, and sharing of health information. Compliance with these regulations is not only a legal obligation but also a fundamental ethical imperative for healthcare providers operating in the digital age. Navigating this complex regulatory landscape requires ongoing vigilance and adaptation to ensure that technological advancements align with the principles of patient privacy and data security.

Navigating Data Privacy and Security Concerns

Avoiding data breaches and mishandling navigating data privacy and security concerns in the context of digital transformation involves several key strategies.

- **Implementing robust security measures**: Any healthcare organization must secure the patient's data. This would include necessary and recommended data protection practices for the organization that are encrypted, strong user authentication, regular security audits, and training staff.

- **Complying with regulations**: The healthcare institution must ensure compliance with the relevant regulations in matters of data privacy and security. It needs a policy and procedure manual that is up-to-date through regular audits and amendments of policies corresponding with the changes in the regulations.

- **Promoting transparency**: Organizations in the healthcare sector should be more forthright in sharing ways through which they collect, use, and secure the data of their patients. This way, they will become responsible and thereby gain the trust of the customers, who are the patients, to be able to share the data with them comfortably.

- **Using advanced technology**: Technology can be a big tool to follow in cases of data privacy and security concerns. For instance, the use of artificial intelligence and machine learning can play a big role in showing the possibility of an anomaly that may point to a data breach.

Now, let's go deeper into the interplay of these strategies with the usability aspect of digital healthcare.

The Convergence of Security and Usability

These are the critical domains where healthcare IT systems must converge if digital healthcare solutions are to be successfully realized: security and usability. Security cannot be an afterthought; it must be integrated into the design from the very beginning. Incorporation of this strict security always leads to complex user interfaces that may, in turn, harm the user experience, especially for those who are not skilled in technology. The difficulty is in devising a user interface that is intuitively navigable yet strong enough to safeguard sensitive data; understanding the possible pitfalls and end-user needs well is therefore necessary. While healthcare professionals need efficient systems that do not disrupt or hinder their workflow, patients must have easy and open interfaces.

One of the techniques realized is the need for role-based access control (RBAC) systems (see `www.paubox.com/blog/what-is-role-based-access-control`). RBAC ensures that users gain access to information and functionality based on their roles. This is more security-sensitive in that it reduces unauthorized access to sensitive data. It is also a good way of simplifying the user interface since the user does not see things irrelevant to their case.

The other is that the security put in place is adaptive in nature. Such systems adapt to the context of use. For instance, in high-risk scenarios, strong forms of authentication are to be taken, while in those considered less risky, streamlined access is still offered. The principle would be to make security not at the cost of usability. Good design should guide the user gently through required security protocols without feeling cumbersome. That is a very important balance of acceptance and efficiency of the digital healthcare system.

User-Centric Design: More Than Just Aesthetics

User-centric design in healthcare IT systems aims to understand and meet the needs and behavioral patterns of the end users. It is about devising systems that are at the same time intuitive and approachable enough to prove useful to a whole range of people, from tech-savvy doctors to older patients unfamiliar with digitization. This design philosophy prescribes that the users should be involved in the development process through conducting surveys, interviews, and usability testing. For example, patient involvement in the design of a patient portal could reveal insight into preferences and problems in setting up a system that is secure, genuinely helpful, and easy to use.

Accessibility and Inclusivity: Toward Universal Usability

With digital transformation, accessibility, and inclusivity are an obvious must for all healthcare systems that are not willing to leave any patient behind. Such designs consider systems that are usable and understandable by people with various forms of disabilities, as well as individuals from different cultural and linguistic backgrounds. These can include a screen reader to translate simple language and can be initiated verbally, like those for the visually impaired. These are not only ethical considerations but enhance the overall security and functionality of the system by making it more usable for a wider population.

Tackling Scalability Challenges

The large-scale implementation of useful and usable healthcare systems poses huge challenges, particularly in reaching an even level of consistency and reliability on all platforms and devices. In essence, scalability embodies the aspects of increasing the magnitude of data and maintaining high uniformity standards in security and usability. Cloud-based solutions offer scalability, which prepares for the development of future technologies that can ensure healthcare systems are capable of handling demand and flexible enough to deal with change.

Insights from Real-World Applications

Real-world applications of a user-friendly and secure system for healthcare can be enlightening. One of the illustrative examples, in this case, is a study from the internal medicine department of a Dutch hospital, reporting barriers to and facilitators of the implementation and navigation of the

A 2021 EHR system study describes the phenomenon of workarounds being used by healthcare professionals during the day-to-day operation of the newly implemented EHR system in a hospital to identify which kind of workarounds remain and their user-perceived consequences (see Figure 10-1). The workarounds were applied in different forms, with most using separate text fields, copying and pasting of data, or bypassing some functionalities of the EHR system. This was applied mostly in a bid to save time or mismatch between the functionalities of the EHR system.

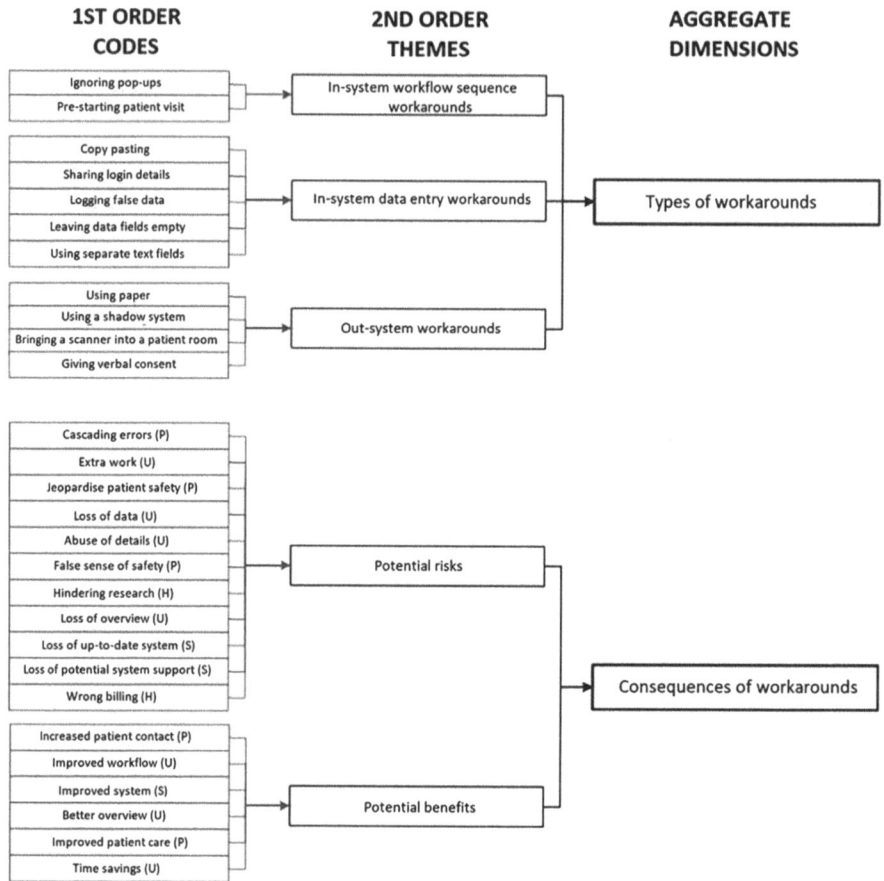

Figure 10-1. *Final data structure: workarounds by healthcare professionals*
Source: *https://bmcmedinformdecismak.biomedcentral.com/articles/10.1186/s12911-021-01548-0/figures/1*

Navigating Future Innovations

The healthcare sector is poised at the forefront of a technological revolution, eagerly embracing bleeding-edge innovations such as AI, blockchain, and the Internet of Medical Things (IoMT). These technologies promise to reshape the future of healthcare, transforming diagnosis, data management, and treatment methodologies.

On its part, AI is transforming patient care through the predictive analytics it brings to personalized medicine. AI analyzes vast datasets, uncovering patterns and insights that may be missed by human clinicians, resulting in more accurate diagnoses and targeted treatment plans. However, the integration of AI also raises sensitive issues concerning data privacy, algorithmic biases, and the ethical use of patient information. AI systems must be transparent and fair and adhere to stringent privacy regulations to safeguard sensitive patient information and maintain trust.

Blockchain technology provides a powerful weapon against the data insecurity and privacy issues prevailing in the healthcare sector. By creating decentralized, tamper-resistant records, blockchain ensures the protection of patient data and facilitates the secure sharing of information among stakeholders without the risk of hacking or misuse. The enhanced traceability and security offered by blockchain technology significantly improve the accountability of medical products and services.

Furthermore, blockchain's immutable ledger can streamline clinical trials, ensuring transparency and trust in research data. Smart contracts, powered by blockchain, can automate processes such as insurance claims and supply chain management, reducing administrative overhead and increasing efficiency. The potential of blockchain to revolutionize healthcare data management is immense, but its successful implementation requires collaboration among stakeholders and the establishment of industry-wide standards.

IoMT is another frontier for extending telemedicine and remote monitoring capabilities. Wearable health monitors and other connected medical devices are empowering proactive healthcare and timely interventions. However, the vast scope of IoMT also introduces vulnerabilities, as each connected device becomes a potential entry point for cyberattacks. Robust security protocols and frequent updates are essential to safeguard IoMT devices from the ever-evolving threat landscape.

The integration of IoMT into healthcare holds the promise of personalized care, early disease detection, and improved chronic disease management. Remote monitoring allows healthcare providers to track patients' health parameters in real time, facilitating timely interventions and reducing hospital readmissions. IoMT-enabled devices can also empower patients to take an active role in managing their health, leading to better outcomes and a higher quality of life.

As technology continues to evolve, proactive adaptation will be key for healthcare organizations to harness the full potential of these innovations. Training healthcare providers on new technologies, ensuring cross-compatibility of systems, and establishing clear guidelines and ethical standards for their use are essential steps. This forward-looking approach will enable the healthcare system to fully leverage the benefits of these technologies for improved patient care while safeguarding privacy and security with the highest rigor.

The future of healthcare lies in the harmonious integration of these innovations, where advanced technology complements human expertise to create a more efficient, effective, and safe patient-centric healthcare system.

By striking the right balance among these elements, the healthcare industry can embrace digital transformation and deliver better, safer, and more efficient care for all. This transformative journey requires collaboration among stakeholders, including healthcare providers, technology companies, regulators, and patients, to ensure the ethical and responsible use of technology for the betterment of healthcare. Only then can the promise of a technologically advanced, patient-centric healthcare system become a reality.

Conclusion

The quest for balance between data privacy, security, and usability is ongoing and dynamic. It's a responsibility we all shoulder—individuals, technology leaders, and legislators alike. Our actions, choices, and policies today shape the digital landscape of tomorrow. We must actively participate in the conversation, demanding transparency, accountability, and safeguards from those who collect and utilize our personal information.

Success means rejecting the notion of inevitable trade-offs. It's not a choice between privacy or progress or security or convenience. We can have both. Innovation should not come at the cost of our fundamental rights. Instead, we must strive to create technologies and regulations that prioritize privacy by design, ensuring that our data is shielded from unauthorized access, breaches, and misuse.

Embracing a future where respect for privacy forms the foundation, innovation can flourish, and data can serve its full potential to improve lives without leaving anyone behind. This means empowering individuals with control over their data, enabling them to make informed decisions about how it is collected, shared, and used. It also means fostering a culture of responsible data practices, where organizations prioritize transparency, ethical use, and robust security measures to build trust and confidence. By working together, we can create a digital ecosystem that benefits everyone, where privacy is a fundamental right that is upheld and respected.

PART IV

Industry Perspective

CHAPTER 11

Success Stories in Software Usability

Usability is one of the most critical factors determining the success of software products. It refers to how easily users can accomplish their goals when interacting with a system. In today's digital-first world, intuitive, seamless experiences are not just a luxury—they are a necessity. Poor usability often leads to frustration, abandonment, and failure of software products, while exceptional usability fosters user satisfaction, loyalty, and widespread adoption.

This chapter delves into 20 real-world examples of how companies across diverse sectors leveraged usability to create successful software products. These examples span industries such as mobile technology, entertainment, e-commerce, financial services, education, and more. Each highlights how usability principles such as simplicity, personalization, performance, and accessibility helped transform software into indispensable tools. These success stories underscore the vital role that usability plays in shaping products that delight users, drive business success, and change industries.

© Saurav Bhattacharya 2025
P. Gujar, *Data Usability in the Enterprise*, https://doi.org/10.1007/979-8-8688-1183-8_11

Apple's iPhone: Revolutionizing Mobile Usability

When Apple launched the iPhone in 2007, it wasn't just introducing a new phone—it was reinventing how people interacted with technology. Prior to the iPhone, mobile phones relied heavily on physical keyboards and complex menus that limited functionality. Apple's approach, rooted in intuitive usability, turned the iPhone into one of the most transformative devices of the twenty-first century.

Usability Breakthroughs

- **Touchscreen interface**: The iPhone's capacitive touchscreen was a radical departure from the hardware keyboards and styluses that dominated the mobile landscape. With simple gestures like tapping, swiping, and pinching, the iPhone eliminated the need for physical input devices and made navigation feel natural and intuitive.

- **Home screen simplicity**: The iPhone's home screen featured a grid of clearly labeled app icons, with a single home button to return users to the main screen. This simplification of navigation minimized confusion and allowed users to access their apps with ease.

- **App Store ecosystem**: By introducing the App Store in 2008, Apple further enhanced the iPhone's usability. Users could download third-party apps with just a few taps, adding new features and functionalities that integrated seamlessly into the phone's interface.

Usability Enhancements over Time

Over the years, Apple continued to refine the iPhone's usability. Features like Siri, Apple's voice-activated assistant, and Face ID, which allows users to unlock their phones with facial recognition, further streamline user interaction. Apple's commitment to user-centered design has enabled the iPhone to maintain its usability leadership in the smartphone market for over a decade.

Impact of Usability

The iPhone's usability revolutionized the mobile industry, making smartphones accessible to people of all technical abilities. Its intuitive design, coupled with its focus on user experience, allowed Apple to dominate the smartphone market and set the standard for mobile usability. By removing the barriers between technology and users, Apple created a product that is easy to use yet incredibly powerful, fundamentally changing the way people engage with mobile technology.

Google Search: Mastering the Power of Simplicity

Since its launch in 1998, Google Search has remained the undisputed leader in online search. A major reason for its dominance is its focus on simplicity and usability. Google succeeded in creating a search engine that is incredibly powerful yet remarkably easy to use, ensuring that users can find information with minimal effort.

Usability Breakthroughs

- **Minimalist design**: At the heart of Google's success is its uncluttered interface. The Google homepage consists of just a search bar and two buttons, eliminating distractions and helping users focus solely on searching for information. This minimalist design became the standard for search engines and has remained largely unchanged for over two decades.

- **Autocomplete and suggestions**: Google introduced autocomplete to speed up the search process. As users type their queries, Google suggests potential search terms based on popular queries and individual search history, allowing users to find what they're looking for more quickly and with fewer keystrokes.

- **Fast performance**: One of the key factors behind Google's usability is its speed. The search engine delivers results almost instantaneously, providing users with quick access to the information they need. This commitment to speed has been a cornerstone of Google's user experience.

Innovations in Search Usability

Over time, Google continued to refine its search experience. The introduction of features like Google Knowledge Graph, which provides instant answers to questions without needing to click links, and voice search, which allows users to conduct searches using spoken queries, further enhanced Google's usability.

Impact of Usability

Google's focus on simplicity, speed, and relevance has made it the most widely used search engine in the world, processing over 3.5 billion searches daily. Its usability innovations have helped solidify Google's position as a trusted source of information and a critical tool for navigating the web. By making search easy, fast, and intuitive, Google has forever changed the way people interact with the Internet.

Netflix: Usability Across Platforms for a Seamless Experience

Netflix started as a DVD rental service but evolved into one of the world's most popular streaming platforms, with over 230 million subscribers worldwide. A significant factor in Netflix's success is its unwavering focus on usability, ensuring a seamless and engaging experience across all devices, from smart TVs and gaming consoles to smartphones and tablets.

Usability Breakthroughs

- **Cross-device consistency**: Netflix's interface is consistent across all devices, providing a familiar experience whether users are watching on a mobile phone, tablet, smart TV, or computer. This consistency reduces the cognitive load on users, allowing them to effortlessly transition between devices without having to relearn how to navigate the interface.

- **Personalized recommendations**: Netflix's recommendation engine is powered by sophisticated algorithms that analyze users' viewing habits and preferences to suggest content tailored to their tastes. This personalized approach enhances usability by helping users discover new content with minimal effort.

- **User-friendly navigation**: Netflix's interface is designed for simplicity, with horizontal scrolling through rows of content organized by genre, recommendations, and watch history. Features like Continue Watching and My List make it easy for users to resume shows or save content for later.

Adaptive Streaming and Performance

Another key to Netflix's usability is its adaptive streaming technology, which adjusts the video quality based on the user's Internet connection. This ensures that users can enjoy a smooth viewing experience regardless of bandwidth fluctuations, enhancing overall usability by reducing buffering and improving video playback.

Impact of Usability

Netflix's focus on cross-device usability, personalized recommendations, and smooth streaming has made it the leader in the streaming industry. Its ability to deliver a consistent, high-quality user experience across multiple platforms has helped it build a loyal global audience, shaping how people consume entertainment in the digital age.

Slack: Enhancing Collaboration with Enterprise Usability

Slack is a workplace collaboration tool that has become indispensable for teams across various industries. Its success lies in its ability to streamline communication, simplify project collaboration, and integrate with other productivity tools, all while maintaining a highly usable interface.

Usability Breakthroughs

- **Channel-based communication**: Slack organizes conversations into channels based on teams, projects, or topics. This structure helps reduce clutter in communication and ensures that discussions remain focused and relevant. Users can easily navigate between channels to stay up to date on specific conversations, increasing productivity and reducing confusion.

- **Searchable message history**: One of Slack's most valuable usability features is its powerful search functionality. Users can quickly search through their message history, files, and links to find specific information, reducing the time spent hunting through email chains or chat logs.

- **Third-party integrations**: Slack's usability is enhanced by its ability to integrate with a wide range of third-party applications, such as Google Drive, Trello, Zoom, and GitHub. These integrations allow users to perform tasks, manage projects, and communicate without leaving the Slack interface, simplifying workflows and improving efficiency.

317

Usability Enhancements for Remote Work

During the COVID-19 pandemic, Slack became even more critical for remote teams. Features like **video and voice calls**, **screen sharing**, and **status updates** helped teams stay connected and productive while working from home. Slack's ability to scale and adapt to the needs of remote workers further solidified its position as a must-have tool for modern businesses.

Impact of Usability

Slack's emphasis on usability, from its intuitive channel structure to its seamless integrations, has made it a favorite collaboration tool for organizations of all sizes. By simplifying workplace communication and making it easier to manage projects, Slack has become an essential tool in the modern workplace, driving efficiency and productivity.

Zoom: Redefining Usability in Video Conferencing

Zoom rose to prominence during the global COVID-19 pandemic, but its success was built on a foundation of usability-focused design. While video conferencing tools like Skype and Microsoft Teams have been around for years, Zoom's intuitive interface, easy access, and reliable performance have made it the platform of choice for businesses, schools, and individuals around the world.

Usability Breakthroughs

- **One-click meeting access**: Zoom's one-click meeting feature allows users to join video calls with a single click without needing to create accounts or download complex software. This ease of access made Zoom especially appealing to users who were unfamiliar with video conferencing technology.

- **Simple interface with key features**: Zoom's interface is designed for ease of use, with key features like muting, screen sharing, and chat accessible with just a few clicks. These essential functions are prominently displayed, allowing users to participate in meetings without feeling overwhelmed by too many options.

- **Breakout rooms for collaboration**: Zoom introduced **breakout rooms**, which allow hosts to split meeting participants into smaller groups for focused discussions. This feature was particularly valuable for educational settings, virtual conferences, and team collaboration, offering a more interactive and personalized experience.

Performance and Reliability

Zoom's performance was another critical factor in its usability. The platform was designed to maintain high-quality video and audio, even in low-bandwidth environments, ensuring that users could rely on it for uninterrupted communication. This reliability made Zoom a go-to platform for virtual meetings, classes, and social gatherings.

Impact of Usability

Zoom's focus on usability, combined with its robust performance, helped it achieve exponential growth during the pandemic. By providing a simple, reliable solution for video conferencing, Zoom became the standard for remote communication, used by millions of people worldwide. Its success demonstrates the importance of making technology accessible and intuitive, especially during times of rapid change and uncertainty.

Amazon: Creating a User-Centered E-Commerce Experience

Amazon has revolutionized the way people shop, and much of its success can be attributed to its focus on usability. By creating a seamless, intuitive shopping experience, Amazon has become the largest online retailer in the world, catering to millions of customers across the globe.

Usability Breakthroughs

- **One-click purchasing**: Amazon's one-click purchase feature is a perfect example of reducing friction in the shopping process. With just a single click, users can complete their purchases, bypassing the traditional multi-step checkout process. This simplicity makes shopping faster and easier, reducing cart abandonment rates and increasing sales.

- **Personalized shopping experience**: Amazon's recommendation engine uses machine learning to suggest products based on users' browsing and purchase history. These personalized recommendations improve usability by helping users discover new products that align with their preferences, making shopping more enjoyable and efficient.

- **Easy returns and customer support**: Usability extends beyond the purchase process, and Amazon's hassle-free return policy and responsive customer support further enhance the user experience. By making it easy for users to return products or resolve issues, Amazon builds trust and encourages repeat purchases.

Usability for Mobile Shopping

With the rise of mobile shopping, Amazon ensured that its app offered the same usability features as its desktop site. The Amazon app allows users to browse, compare, and purchase products with ease, offering a consistent experience across devices. Features like **barcode scanning** and **voice shopping** (through Alexa) further enhance usability on mobile devices.

Impact of Usability

Amazon's relentless focus on usability has helped it build a massive global customer base. By simplifying the shopping process, offering personalized recommendations, and providing excellent post-purchase support, Amazon has set new standards for e-commerce usability. Its user-centered approach has transformed the way people shop online, making it the go-to platform for millions of consumers.

Microsoft Teams: Usability in a Remote Collaboration World

Microsoft Teams became a critical tool for businesses and educational institutions during the COVID-19 pandemic, offering a platform for remote collaboration, video conferencing, and file sharing. Teams' success lies in its ability to provide a unified, highly usable experience for distributed workforces.

Usability Breakthroughs

- **Seamless integration with Microsoft Office**: One of the key factors behind Microsoft Teams' usability is its integration with the Microsoft Office suite, including Word, Excel, and PowerPoint. This integration allows users to collaborate on documents in real-time without leaving the Teams environment, streamlining workflows and improving productivity.

- **Customizable channels**: Teams allows users to create dedicated channels for different projects, departments, or initiatives, ensuring that communication is organized and focused. By keeping conversations within specific channels, Teams reduces information overload and helps users stay on top of relevant discussions.

- **Cross-platform usability**: Teams is available on desktop, mobile, and web platforms, providing a consistent user experience across devices. This flexibility is essential for remote workers who need to stay connected and productive from various locations and devices.

Enhancements for Remote Work

As remote work became the norm during the pandemic, Microsoft Teams introduced features like Together Mode, which placed participants in a shared virtual environment to mimic in-person meetings, and background effects, which allowed users to customize their video backgrounds. These features enhanced usability by creating a more engaging and personalized remote meeting experience.

Impact of Usability

Microsoft Teams' focus on usability made it a vital tool for remote work during the pandemic. Its integration with Microsoft Office, customizable channels, and cross-platform availability allowed organizations to stay connected and productive in a distributed work environment. Teams' usability has helped it become one of the most widely used collaboration platforms globally.

Duolingo: Making Language Learning Usable and Fun

Duolingo has transformed the way people learn new languages by offering a fun, gamified approach to language acquisition. Its success lies in its ability to make learning accessible to users of all ages and skill levels while maintaining a high degree of usability.

Usability Breakthroughs

- **Gamification for engagement**: Duolingo uses game mechanics like points, levels, streaks, and rewards to keep users motivated and engaged. These features make learning feel like a game, encouraging users to return to the app regularly and complete lessons.

- **Short, manageable lessons**: Duolingo's lessons
  are designed to be completed in just a few minutes,
  allowing users to fit learning into their daily routines.
  This bite-sized approach makes the app easy to use and
  reduces cognitive load, helping users stay focused and
  retain information more effectively.

- **Adaptive learning path**: Duolingo's platform adapts to
  the user's progress, offering more challenging exercises
  as the user improves. This personalized learning path
  ensures that users are always working at an appropriate
  level, enhancing usability by providing a customized
  learning experience.

Expanding Usability for Language Learning

Duolingo expanded its usability features by introducing Duolingo for
Schools, a platform that allows teachers to track their students' progress,
assign lessons, and provide feedback. This expansion into the educational
space further enhanced Duolingo's usability by catering to both individual
learners and educators.

Impact of Usability

Duolingo's user-centered design and gamified approach have made it one
of the most popular language-learning platforms in the world, with over
500 million users. By making language learning fun, accessible, and easy
to fit into daily life, Duolingo has revolutionized the way people acquire
new languages, empowering users across the globe to achieve their
language goals.

Dropbox: Simplifying Cloud Storage Usability

Dropbox is one of the pioneers of cloud storage, offering a simple and intuitive way for users to store, sync, and share files across multiple devices. Dropbox's focus on usability helped it become a leader in the cloud storage industry.

Usability Breakthroughs

- **Drag-and-drop interface**: Dropbox's drag-and-drop functionality made file storage and management incredibly simple. Users could upload files by simply dragging them into the Dropbox folder without needing to navigate complex menus or configurations.

- **Seamless syncing across devices**: Dropbox's ability to sync files across multiple devices automatically was a key feature that set it apart from traditional file storage solutions. Users could access their files from any device, and changes made on one device were instantly reflected across all others.

- **Sharing and collaboration**: Dropbox made it easy to share files and collaborate with others, even if the recipients didn't have a Dropbox account. This simplicity removed barriers to collaboration and made Dropbox an essential tool for both personal and professional use.

Enhancing Usability with Business Features

Dropbox later expanded its offerings with Dropbox Business, which added features like team management, advanced security controls, and file recovery. These enhancements maintained the platform's core usability while catering to the needs of larger organizations and teams.

Impact of Usability

Dropbox's focus on simplicity and ease of use helped it become one of the most trusted names in cloud storage. By offering an intuitive, reliable solution for file storage and sharing, Dropbox changed the way individuals and businesses think about file management, making cloud storage an everyday part of modern life.

Uber: Streamlining the Transportation Experience

Uber revolutionized the transportation industry by creating an app that made hailing a ride as simple as tapping a button. Its usability-focused approach transformed how people get around, offering a more convenient alternative to traditional taxis and public transportation.

Usability Breakthroughs

- **One-tap ride requests**: Uber's app allows users to request a ride with a single tap. The app's integration with GPS enables it to automatically detect the user's location, eliminating the need for users to input their pickup address manually. This simplicity makes the process of booking a ride incredibly fast and convenient.

- **Transparent pricing**: Uber provides fare estimates before users confirm their ride, ensuring transparency and preventing price surprises. Users can compare fare options, such as UberX, Uber Pool, and Uber Black, based on their budget and preferences.

- **Seamless payment integration**: Uber integrates payment directly into the app, allowing users to store their payment information and handle transactions without needing cash. After completing the ride, payment is automatically processed, and users can rate their driver—streamlining the entire process.

Expanding Usability with New Services

Uber expanded its usability by introducing services like **Uber Eats**, which allows users to order food for delivery with the same seamless experience as booking a ride. The integration of these additional services further solidified Uber's position as a usability-driven platform.

Impact of Usability

Uber's emphasis on usability transformed the way people think about transportation. By making the process of booking a ride fast, easy, and transparent, Uber became the global leader in ride-hailing. Its usability-first approach helped Uber expand rapidly and set the standard for on-demand transportation services worldwide.

Spotify: Delivering Personalized Usability in Music Streaming

Spotify is one of the world's leading music streaming platforms, with over 500 million active users. Its success is rooted in its ability to provide users with a personalized and intuitive music listening experience, ensuring that they can easily discover and enjoy new music.

Usability Breakthroughs

- **Personalized playlists**: Spotify's **Discover Weekly** and **Daily Mix** playlists are automatically generated based on the user's listening habits. These playlists help users discover new music that aligns with their tastes, making the music discovery process effortless and enjoyable.

- **Seamless device syncing**: Spotify's **cross-device syncing** feature allows users to switch between devices without interrupting their listening experience. Whether a user starts a playlist on their phone and continues it on their desktop, the transition is seamless, enhancing usability across platforms.

- **Offline listening**: Spotify's offline mode enables users to download playlists and albums for listening without an Internet connection. This feature is particularly useful for users in areas with limited connectivity or during travel, making the platform more usable in a wider range of situations.

Expanding Usability with Podcasts

Spotify has expanded its usability by becoming a major player in the podcasting space. By integrating podcasts directly into its platform, Spotify allows users to discover and listen to podcasts with the same ease as music, making it a one-stop shop for all audio content.

Impact of Usability

Spotify's focus on personalized usability and seamless cross-device integration has made it the go-to platform for music lovers around the world. By making it easy for users to discover new music, create personalized playlists, and enjoy content across devices, Spotify has redefined the way people engage with music in the digital age.

Airbnb: Redefining Travel with a Focus on Usability

Airbnb disrupted the traditional hospitality industry by offering travelers an easy-to-use platform for booking unique accommodations, from city apartments to country cabins. Its success is largely due to its focus on creating a seamless and user-friendly experience for both hosts and guests.

Usability Breakthroughs

- **Intuitive search and booking**: Airbnb's search and booking process is designed to be as simple as possible. Users can filter search results based on price, location, amenities, and other criteria to find accommodations that meet their needs. The booking process is streamlined, requiring just a few clicks to secure a reservation.

329

- **High-quality photos and reviews**: Airbnb emphasizes high-quality photos and detailed guest reviews, giving users the information they need to make informed decisions about their stay. This transparency builds trust between hosts and guests and enhances the overall user experience.

- **Mobile app usability**: Airbnb's mobile app offers the same functionality as its web platform, allowing users to search for and book accommodations on the go. The app is optimized for mobile use, ensuring a consistent experience across devices.

Expanding Usability with Experiences

In addition to accommodation bookings, Airbnb expanded its offerings with Airbnb Experiences, which allows travelers to book activities hosted by locals. This expansion adds to Airbnb's usability by providing a comprehensive platform for both accommodation and activities, all in one place.

Impact of Usability

Airbnb's focus on usability helped it become one of the most popular travel platforms in the world, connecting millions of travelers with unique and personalized accommodations. By simplifying the booking process and building trust through transparency, Airbnb has redefined how people travel and explore new destinations.

PayPal: Trust and Security in Payment Usability

PayPal is a global leader in online payments, known for its secure and easy-to-use platform. Its focus on usability and trust has made it a preferred choice for millions of users worldwide.

Usability Breakthroughs

- **Simple payment process**: PayPal's payment process is straightforward and user-friendly. Users can send and receive money using just an email address, eliminating the need for complicated bank transfers. This simplicity makes PayPal accessible to users of all ages and technical abilities.

- **Buyer protection and security**: PayPal offers buyer protection, ensuring that users feel secure when making purchases online. By providing refunds for unauthorized transactions and protecting against fraud, PayPal builds trust and encourages users to make purchases with confidence.

- **Mobile app convenience**: PayPal's mobile app offers the same functionality as its web platform, allowing users to send money, pay bills, and make online purchases with just a few taps. The app's user-friendly design ensures that users can complete transactions quickly and efficiently, even on the go.

Expanding Usability with Venmo

PayPal expanded its usability offerings by acquiring **Venmo**, a peer-to-peer payment platform known for its social features and ease of use. Venmo allows users to send money to friends and family with just a few taps, making it ideal for splitting bills or sharing expenses.

Impact of Usability

PayPal's focus on simplicity, security, and user trust has made it one of the most popular online payment platforms in the world. By making it easy for users to send and receive money securely, PayPal has built a loyal customer base and helped shape the future of digital payments.

TikTok: Short-Form Video, Long-Term Usability Success

TikTok has taken the world by storm, emerging as one of the most popular social media platforms with over one billion active users. The app's usability, driven by its intuitive design and personalized content discovery, has been a key factor in its rapid growth and global appeal.

Usability Breakthroughs

- **Infinite scroll for content discovery**: TikTok's For You page presents users with an endless stream of videos tailored to their interests. The infinite scroll design, which allows users to swipe up to view new content continuously, makes content discovery effortless and engaging.

- **Simple video creation tools**: TikTok's built-in video editing tools are designed to be easy for anyone to use, regardless of technical skill. Users can add filters, music, text, and effects with just a few taps, making it simple to create professional-looking content in seconds.

- **Personalized algorithm**: TikTok's algorithm learns from users' interactions—such as likes, shares, and watch time—to deliver personalized content. This personalization keeps users engaged by showing them videos that align with their interests, enhancing the platform's usability.

Enhancing Usability with Trends and Challenges

TikTok leverages hashtags, trends, and challenges to keep users engaged and inspire content creation. These features encourage users to participate in viral trends, making the platform more interactive and community-driven, which further enhances usability.

Impact of Usability

TikTok's focus on usability, personalization, and simplicity has made it one of the fastest-growing social media platforms in the world. Its intuitive design and engaging content discovery experience have captured the attention of millions, transforming how people consume and create short-orm video content.

Shopify: Usability for Small Businesses and Entrepreneurs

Shopify is an e-commerce platform that empowers small businesses and entrepreneurs to build and manage their online stores. Shopify's success is built on its focus on usability, providing users with an easy-to-use platform to create professional e-commerce websites without requiring technical expertise.

Usability Breakthroughs

- **Easy store setup**: Shopify simplifies the process of setting up an online store. With pre-designed themes and drag-and-drop functionality, users can create a professional-looking website in minutes, even if they have no coding experience.

- **Comprehensive dashboard**: Shopify's dashboard provides users with an intuitive overview of their store's performance, including sales, orders, and customer data. The dashboard's user-friendly design makes it easy for store owners to manage their business and make informed decisions.

- **App integrations**: Shopify's app store offers a wide range of third-party integrations, allowing users to add features like email marketing, shipping solutions, and payment gateways with just a few clicks. These integrations enhance the platform's usability by giving store owners the flexibility to customize their store to meet their needs.

Expanding Usability with Shopify Payments

Shopify introduced Shopify Payments, its payment processing solution, to further simplify the e-commerce experience. By integrating payments directly into the platform, Shopify eliminates the need for third-party payment gateways, streamlining the checkout process for both merchants and customers.

Impact of Usability

Shopify's focus on usability has made it the platform of choice for millions of small businesses and entrepreneurs. By offering an easy-to-use solution for building and managing online stores, Shopify has democratized e-commerce and empowered individuals to succeed in the digital marketplace.

Figma: Usability and Real-Time Collaboration in Design

Figma is a web-based design tool that has revolutionized the way teams collaborate on design projects. Its focus on real-time collaboration and usability has made it a favorite among designers, developers, and product teams.

Usability Breakthroughs

- **Real-time collaboration**: Figma allows multiple users to work on the same design file simultaneously, similar to how Google Docs enables collaborative editing. This real-time collaboration feature eliminates the need for back-and-forth file sharing and makes it easier for teams to work together on design projects.

335

- **Cloud-based platform**: As a cloud-based design tool, Figma allows users to access their design files from any device with an Internet connection. This flexibility enhances usability by ensuring that users can work from anywhere without needing to install software or worry about version control.

- **Component libraries**: Figma's **component libraries** allow users to create reusable design elements that can be shared across projects. This feature improves usability by streamlining the design process and ensuring consistency across teams.

Expanding Usability with Design Systems

Figma's support for design systems allows organizations to create standardized design libraries that can be used across products and teams. This feature enhances usability by promoting consistency in design, making it easier for teams to collaborate on large-scale projects.

Impact of Usability

Figma's focus on real-time collaboration, cloud accessibility, and design system management has made it one of the most popular design tools on the market. By making it easy for teams to collaborate and create consistent designs, Figma has redefined how design work is done in the digital age.

Canva: Democratizing Design with Usability

Canva is an online graphic design platform that enables users to create professional-quality designs with ease. By prioritizing usability, Canva has made design accessible to non-designers, empowering individuals and businesses to create high-quality visuals without needing advanced technical skills.

Usability Breakthroughs

- **Drag-and-drop design**: Canva's drag-and-drop interface makes it easy for users to create designs by selecting templates, images, and elements and arranging them on the canvas. This simplicity allows users to create polished designs without needing to learn complex design software.

- **Templates for every occasion**: Canva offers thousands of pre-designed templates for everything from social media posts to business cards and presentations. These templates make it easy for users to get started quickly and customize their designs to suit their needs.

- **Collaborative design features**: Canva allows users to collaborate on designs in real time, making it easier for teams to work together on projects. This feature enhances usability by streamlining the design process and enabling feedback and iteration without the need for multiple versions of the same file.

Expanding Usability with Canva Pro

Canva introduced Canva Pro, a premium offering that provides additional features such as brand kits, custom fonts, and advanced design tools. These features enhance usability for businesses and professional designers, making it easier to create consistent, on-brand designs at scale.

Impact of Usability

Canva's focus on usability has democratized design, allowing millions of people to create high-quality visuals without needing formal design training. By offering an intuitive platform with powerful features, Canva has become the go-to tool for individuals, small businesses, and marketing teams looking to create professional designs quickly and easily.

Robinhood: Usability in Financial Technology for First-Time Investors

Robinhood disrupted the financial industry by making stock trading accessible to first-time investors through its easy-to-use app. Its focus on usability and simplicity has helped it attract millions of users, particularly among younger, tech-savvy individuals.

Usability Breakthroughs

- **Commission-free trading**: Robinhood introduced commission-free trading, which removed a significant barrier to entry for first-time investors. By eliminating fees, Robinhood made investing more accessible and appealing to a broader audience.

- **Simple, clean interface**: Robinhood's app features a minimalist design, with clear navigation and intuitive controls. Users can easily buy and sell stocks, monitor their portfolios, and access market data, all from a single dashboard.

- **Fractional shares**: Robinhood introduced fractional shares, allowing users to invest in high-priced stocks with as little as $1. This feature makes investing more accessible to people who may not have large amounts of capital to invest.

Expanding Usability with Robinhood Crypto

Robinhood expanded its usability by introducing Robinhood Crypto, allowing users to buy and sell cryptocurrencies alongside stocks. This integration made it easy for users to manage all of their investments in one app, enhancing the overall user experience.

Impact of Usability

Robinhood's focus on usability and accessibility has transformed the investment landscape, making it easier for first-time investors to participate in the stock market. By removing barriers to entry and offering a simple, intuitive platform, Robinhood has attracted millions of users and reshaped the future of investing.

Trello: Task Management with Usability-Driven Simplicity

Trello is a task management and collaboration tool that uses a visual, card-based system to help teams organize their work. Trello's focus on simplicity and usability has made it a popular choice for individuals and teams looking for an easy way to manage projects.

Usability Breakthroughs

- **Card-and-board system**: Trello's card-and-board system provides a visual way to manage tasks. Users can create cards for individual tasks and organize them into boards, making it easy to see the status of a project at a glance.

- **Drag-and-drop functionality**: Trello's drag-and-drop functionality allows users to move cards between columns, representing different stages of a project (e.g., To Do, In Progress, Completed). This simplicity makes task management intuitive and user-friendly.

- **Collaboration and Integration**: Trello supports real-time collaboration, allowing team members to comment on cards, assign tasks, and track progress. Trello also integrates with other tools like Slack, Google Drive, and Dropbox, enhancing usability by making it a central hub for project management.

Expanding Usability with Trello Business Class

Trello introduced Business Class, a premium offering with additional features like board templates, custom fields, and advanced reporting. These features enhance usability for larger teams and organizations, making it easier to manage complex projects.

Impact of Usability

Trello's emphasis on simplicity and visual task management has made it one of the most popular project management tools on the market. By offering an easy-to-use platform that works for individuals and teams alike, Trello has helped millions of users stay organized and productive.

Peloton: Merging Usability with Fitness Technology

Peloton is a fitness platform that combines high-quality exercise equipment with interactive, live-streamed fitness classes. Peloton's focus on usability, both in terms of its equipment and its digital platform, has made it a leader in the home fitness industry.

Usability Breakthroughs

- **Intuitive touchscreen interface**: Peloton's stationary bike and treadmill feature large, built-in touchscreens that allow users to browse and join classes with just a few taps. The touchscreen interface is designed for ease of use, making it simple for users to find the workout that fits their schedule and preferences.

- **Live and on-demand classes:** Peloton offers both live-streamed and on-demand classes, giving users the flexibility to work out at their convenience. The platform's intuitive design makes it easy to filter classes by length, instructor, and difficulty level, enhancing the overall user experience.

- **Real-time performance metrics**: Peloton provides users with real-time metrics, such as heart rate, resistance, and cadence, allowing them to track their progress throughout the workout. This feedback helps users stay motivated and engaged, contributing to the platform's usability.

Expanding Usability with the Peloton App

Peloton expanded its usability with the Peloton App, which allows users to access workouts from their mobile devices, even if they don't own Peloton equipment. This expansion makes Peloton's high-quality fitness content available to a wider audience, enhancing usability by offering more flexibility in how users engage with the platform.

Impact of Usability

Peloton's focus on usability, from its intuitive touchscreens to its real-time performance tracking, has made it a leader in the home fitness industry. By combining cutting-edge technology with an engaging, user-friendly platform, Peloton has transformed how people work out, offering a convenient and motivating fitness experience that millions of users have embraced.

Conclusion

The 20 examples presented in this chapter highlight how companies across diverse industries have leveraged usability to create software products that revolutionized their respective markets. By prioritizing user experience, these companies have not only met the needs of their customers but also exceeded expectations, creating products that are intuitive, efficient, and a pleasure to use.

Key Usability Lessons from Success Stories

- **Simplicity is powerful.** From Google's minimalist search interface to Uber's one-tap ride request, simplicity helps reduce friction and enhances user satisfaction.

- **Cross-platform consistency is essential.** Products like Netflix and Spotify provide consistent experiences across devices, ensuring that users can seamlessly switch between platforms without losing functionality.

- **Personalization enhances engagement.** Platforms like Amazon, Netflix, and Spotify leverage personalized recommendations to improve usability, helping users discover content that aligns with their preferences.

- **Reliability and performance matter.** Zoom's ability to provide high-quality video conferencing under low-bandwidth conditions and Google's fast search results show that usability is closely tied to system performance.

- **User-centered design leads to success.** Across all of
  these examples, the companies succeeded by focusing
  on the needs, preferences, and behaviors of their users,
  ensuring that their products are designed with the user
  in mind.

In conclusion, these usability success stories demonstrate that a
commitment to creating intuitive, user-friendly software can drive massive
business success. Whether through simplifying complex processes,
personalizing experiences, or ensuring cross-device usability, these
companies have shown that great design leads to great products—products
that users love, rely on, and recommend to others.

CHAPTER 12

Future Trends in Software Usability and Innovation

The role of usability in software design has evolved dramatically, driven by technological advancements and a growing emphasis on user experience. In the early stages of computing, usability was secondary to functionality, often requiring users to adapt to the limitations of the software. Over time, the focus has shifted, with the understanding that ease of use is central to the adoption and success of software applications.

As we move into an era dominated by emerging technologies, the landscape of usability is undergoing a radical transformation. These innovations are not just changing how users interact with software but are also redefining what is possible in terms of personalized, context-aware, and adaptive user experiences. The following are examples.

- **Artificial intelligence** is enabling software to learn user preferences and personalize interfaces dynamically. Imagine a news app that automatically curates content based on your reading habits or a productivity tool that prioritizes tasks based on your work patterns.

© Saurav Bhattacharya 2025
P. Gujar, *Data Usability in the Enterprise*, https://doi.org/10.1007/979-8-8688-1183-8_12

- **Augmented and virtual reality** are blurring the lines between the physical and digital worlds, creating immersive experiences that demand new approaches to interaction design. Think of surgeons using augmented reality overlays during surgery for enhanced precision or architects walking clients through virtual building models.

- **Voice interfaces** are changing how we interact with devices, allowing for hands-free control and more natural communication. Picture controlling your smart home with voice commands or getting directions while driving without taking your eyes off the road.

- **The Internet of Things** is connecting everyday objects to the Internet, creating a network of intelligent devices that can respond to our needs. Envision a refrigerator that automatically orders groceries when supplies are low or a fitness tracker that provides personalized workout recommendations.

This chapter explores the future trends in software usability and innovation, examining how these advancements in technology are influencing design approaches and user interactions. It investigates the impact of AI on interface personalization, the rise of voice-controlled and gesture-based systems, and the increasing importance of ethical considerations in design. The future of usability lies in creating systems that are intuitive, seamless, and capable of adapting to the needs and contexts of diverse users while maintaining a strong emphasis on accessibility and inclusivity.

The Evolution of Software Usability

From Command Lines to Graphical User Interfaces

In the early days of computing, interactions with software were predominantly text-based, requiring users to input commands in command-line interfaces (CLIs). These systems were efficient for those with technical expertise but inaccessible to the general population. The introduction of graphical user interfaces (GUIs) in the 1980s, popularized by companies such as Apple and Microsoft, revolutionized usability by making software more visually intuitive and easier to navigate.

GUIs introduced elements such as windows, icons, menus, and pointers (often referred to as the WIMP interface), which made software more accessible to non-technical users. This shift marked the beginning of usability as a critical aspect of software development, emphasizing ease of use, visual clarity, and intuitive navigation.

The Web Era and Mobile Usability

With the rise of the Internet in the 1990s, web usability became a focal point of software design. Websites and web applications needed to be simple and efficient to accommodate users with varying levels of expertise. This period saw the emergence of user-centered design (UCD) practices, where user research and usability testing became integral to the design process.

The next major leap in usability came with the advent of smartphones and mobile apps. Mobile usability introduced new challenges, including limited screen real estate, touch-based interactions, and the need for responsive design. Simplicity, speed, and responsiveness became paramount, leading to the development of flat design, minimalist interfaces, and gesture-based interactions that have come to define mobile usability.

347

Usability in the Age of Emerging Technologies

Today, as we transition into an era defined by emerging technologies, usability is once again being redefined. These technologies enable richer, more immersive experiences but also introduce complexity in how users interact with software. In this new landscape, usability is not just about ease of use but also about adaptability, personalization, and creating seamless cross-device experiences.

Artificial Intelligence and Machine Learning in Usability

Artificial intelligence (AI) and machine learning are rapidly transforming how users interact with software, making systems more intelligent, adaptive, and personalized. These technologies enable software to learn from user behavior, predict needs, and offer tailored experiences, significantly enhancing usability.

AI-Driven Personalization

Personalization has become one of the most powerful applications of AI in usability. AI-powered systems analyze user behavior, preferences, and interactions to deliver personalized content, recommendations, and user interfaces. For example, AI can track which features a user frequently accesses and adjust the interface to prioritize those functions, reducing the cognitive load and making the software more intuitive.

Personalization extends beyond content recommendations to interface layout, color schemes, and even interaction modes. By learning user preferences, AI systems can dynamically adapt the user interface to match the user's habits and needs. This results in a more efficient and satisfying user experience, as users spend less time searching for tools or adjusting settings.

AI-driven personalization is particularly valuable in complex applications where users may have different goals and workflows. For example, in an enterprise resource planning (ERP) system, AI can learn the specific tasks an employee performs regularly and customize the dashboard to display the most relevant information and tools.

Anticipatory Design and Predictive Interfaces

Beyond personalizing interfaces, AI is enabling anticipatory design, where software predicts the user's next action and offers proactive assistance. This can manifest in various ways, such as suggesting shortcuts, automating repetitive tasks, or offering context-aware recommendations.

For instance, email applications like Google's Gmail use machine learning to suggest responses or prioritize messages based on the content and past interactions. This reduces the time users spend sorting through emails and improves overall productivity.

As AI becomes more advanced, predictive interfaces will evolve to anticipate complex user needs, enabling users to complete tasks with minimal effort. These systems will be capable of understanding context, intent, and even emotional states, allowing for more seamless and intuitive interactions.

AI for Accessibility

AI also plays a crucial role in improving accessibility. Machine learning algorithms can assist users with disabilities by providing real-time transcription for the hearing impaired, voice commands for those with motor impairments and even image recognition for the visually impaired. AI can help break down barriers to technology, making software usable for a broader range of individuals.

Future AI systems may also be able to adapt interfaces to different levels of cognitive ability, offering simplified interfaces or additional guidance for users who may struggle with complex software.

Natural Language Processing and Voice Interfaces

Natural language processing (NLP) is at the heart of voice interfaces, which are becoming an increasingly popular method of interacting with software. Voice assistants like Amazon Alexa, Google Assistant, and Apple's Siri are paving the way for a more conversational and hands-free user experience.

The Rise of Voice-First Interfaces

Voice-first interfaces represent a shift away from traditional text or touch-based interactions. These interfaces allow users to interact with software using natural language commands, making technology more accessible, especially for users with limited technical skills or physical disabilities.

Voice interfaces are particularly useful in situations where hands-free interaction is necessary, such as driving, cooking, or operating machinery. For instance, voice commands can enable users to control smart home devices, navigate through applications, or retrieve information without the need to use a touchscreen or keyboard.

Conversational Interfaces and Chatbots

NLP also enables conversational interfaces, which include both voice and text-based interactions. These interfaces allow users to interact with software in a more human-like way, using natural language to ask questions, make requests, or perform tasks. Chatbots, for example, are becoming increasingly common in customer service, providing instant responses to user inquiries without the need for human intervention.

The future of conversational interfaces will be defined by their ability to understand and respond to increasingly complex queries. As NLP algorithms become more sophisticated, these interfaces will be able to handle more nuanced and context-aware conversations, allowing users to accomplish tasks more efficiently.

Challenges and Future Directions

Despite their growing popularity, voice interfaces face several challenges that must be addressed to improve usability. One of the main issues is accuracy—NLP systems must be able to understand different accents, dialects, and speech patterns to provide accurate responses. Additionally, these systems must handle ambiguous or vague commands gracefully, offering clarification or suggestions when necessary.

Looking ahead, advancements in NLP and AI will improve the accuracy and reliability of voice interfaces, making them a more integral part of everyday software interactions. We can expect future voice interfaces to support more complex tasks, such as managing multi-step workflows or providing detailed explanations of actions.

Augmented Reality and Virtual Reality Interfaces

Augmented reality (AR) and virtual reality (VR) technologies are set to transform the way users interact with software, offering immersive experiences that blend digital and physical environments. These technologies present new challenges and opportunities for usability design.

Usability in AR Systems

AR overlays digital information onto the real world, enhancing the user's perception of their environment. In terms of usability, AR systems must ensure that digital elements are contextually relevant and easy to interact with. For example, AR navigation apps can display real-time directions overlaid onto the user's surroundings, making it easier to find specific locations.

Designing AR systems requires careful consideration of spatial awareness and user interaction. Virtual elements must be positioned in a way that makes sense in the physical world, and users should be able to manipulate these elements naturally. Usability in AR also depends on the accuracy of motion tracking and the responsiveness of the system to changes in the user's environment.

Immersive VR Experiences

In contrast to augmented reality, virtual reality creates fully immersive digital environments that users can explore and interact with. Usability in VR systems is primarily concerned with creating intuitive, realistic interactions that mimic the physical world. For instance, users should be able to pick up virtual objects, move around within the environment, and interact with virtual characters or interfaces using natural gestures or motions.

One of the key challenges in VR usability is minimizing motion sickness, which can occur when there is a disconnect between the user's physical movements and what they see in the virtual world. Designers must ensure that motion tracking is accurate and that the system responds to user input in real time to create a smooth, immersive experience.

Future Potential and Challenges

As AR and VR technologies continue to evolve, their usability becomes increasingly important. Designers need to focus on creating interfaces that are both intuitive and context-aware, allowing users to navigate and interact with complex virtual environments without confusion or frustration.

The future of AR/VR usability may also include advancements in haptic feedback, allowing users to "feel" virtual objects and adding a new dimension to immersive experiences. Additionally, improvements in motion tracking and AI will enhance the realism and interactivity of these environments, making them more usable for a wider range of applications.

Gesture-Based Interaction

Gesture-based interaction is becoming a key component of modern software usability, especially in contexts where touch or voice input may not be ideal. Gesture control allows users to perform actions by moving their hands, fingers, or body in specific ways, providing a more natural and intuitive method of interaction.

Expanding Use Cases for Gesture Control

Gesture control is particularly valuable in environments where hands-free operation is essential. In healthcare, for example, surgeons can use gesture controls to manipulate medical images without touching any surfaces, reducing the risk of contamination. Similarly, in industrial settings, workers can control machinery or navigate software interfaces using gestures without the need to stop and use a physical input device.

Gesture-based interaction is also gaining traction in consumer electronics. Devices like Microsoft's Kinect, the Leap Motion controller, and modern smartphones with advanced cameras and 3D sensors have

353

popularized gesture control in gaming, entertainment, and everyday applications. For instance, users can swipe their hands to navigate through menus or pinch to zoom on a virtual object.

Overcoming Challenges in Gesture Usability

Despite its potential, gesture-based interaction presents several usability challenges. For gestures to be effective, they must be easy to learn, remember, and perform. Users should not need extensive training to understand how to use gestures effectively, and the system must be able to recognize gestures accurately.

Precision is also a major concern. Inaccurate gesture recognition can lead to user frustration, especially if the system interprets unintended movements as commands. Designers must focus on creating gesture systems that are responsive and reliable, ensuring that the user's actions are recognized and executed correctly.

To address these challenges, future gesture-based systems will likely incorporate more advanced machine learning algorithms and sensor technologies. These improvements should enable more accurate gesture recognition, making gesture control a more viable option for a wider range of applications.

Personalized and Adaptive Interfaces

Personalization and adaptation are becoming fundamental aspects of modern software usability. Users expect software to understand their preferences and behaviors, offering tailored experiences that make interactions more efficient and enjoyable. Personalized and adaptive interfaces take this concept a step further by dynamically adjusting to the user's needs and context.

Adaptive User Interfaces

Adaptive interfaces use AI and machine learning to analyze user behavior and adjust the layout, content, and features accordingly. These systems learn from the user's actions, offering personalized shortcuts, reorganizing menus, and prioritizing frequently used features.

For example, a photo editing application could adapt its interface based on the tools a user frequently uses, simplifying the layout to highlight those tools and minimize distractions. Similarly, a news app might adjust its content feed to prioritize articles that align with the user's reading preferences, providing a more relevant and engaging experience.

Adaptive interfaces reduce the cognitive load on users by eliminating unnecessary steps and streamlining their interactions. However, designers must ensure that these adaptations do not reduce the user's sense of control. Users should be able to customize or override automatic changes if desired, ensuring that the system remains flexible and user-friendly.

Contextual Adaptation

Beyond personalizing the interface based on user preferences, future trends in usability may focus on dynamically adapting the interface to the user's current context. Contextual adaptation considers factors such as the user's location, time of day, activity, and device. For example, a navigation app might switch to a minimal, voice-driven interface when the user is driving, or a fitness app might adjust its interface based on whether the user is indoors or outdoors.

Context-aware interfaces create more seamless and intuitive experiences by anticipating the user's needs and adjusting accordingly. As AI and machine learning technologies continue to advance, the ability of software to understand and respond to user context will improve, enabling more dynamic and personalized interactions.

Cross-Platform Usability and Seamless Experiences

With the increasing number of devices users interact with daily—ranging from smartphones and tablets to wearables, smart TVs, and voice assistants—the demand for seamless, cross-platform experiences has never been higher. Users expect to be able to switch between devices effortlessly, continuing tasks without interruption or the need for repeated logins or configurations.

Importance of Consistency Across Platforms

Cross-platform usability emphasizes the need for consistent user experiences across different devices and operating systems. Whether users are accessing an app on their phone, desktop, or smartwatch, they should encounter a familiar interface that allows them to accomplish tasks without confusion.

Achieving consistency across platforms requires more than just replicating interfaces. Designers must consider the unique constraints and strengths of each device. For instance, a mobile app may need a simplified interface due to limited screen size, while a desktop version may offer more detailed features and functionalities.

Consistency also extends to user data and preferences. Cross-platform applications should sync user settings, preferences, and data in real time, ensuring that users can pick up where they left off, regardless of the device they are using. This creates a more cohesive and integrated experience, reducing frustration and improving usability.

Responsive and Adaptive Design

Responsive design, which adapts the layout and content of an interface based on the device's screen size and capabilities, has become a standard practice in web and mobile development. However, the future of responsive design will go beyond simple screen adjustments. Adaptive design will consider not just the device but also the user's behavior, location, and preferences.

For example, a weather app might offer different levels of detail depending on the device—a glanceable summary on a smartwatch, a detailed forecast on a smartphone, and an interactive weather map on a desktop. By tailoring the content and interface to the device and context, adaptive design creates more usable and relevant experiences.

Cross-platform usability will also be driven by advancements in cloud computing, which allow applications to sync data and preferences across devices in real time. This ensures that users can move seamlessly between devices without losing progress, whether they are playing a game, editing a document, or managing tasks.

Emotional Design and Affective Computing

As software becomes more ingrained in daily life, designers are increasingly recognizing the importance of emotional design—creating interfaces that evoke positive emotions and engage users on a deeper level. Affective computing, the field of study focused on systems that can recognize, interpret, and respond to human emotions, is set to play a central role in future usability innovations.

Designing for Emotional Engagement

Emotional design involves crafting user experiences that resonate emotionally, leading to feelings of satisfaction, trust, or joy. This can be achieved through thoughtful use of color, typography, animation, and interaction feedback. For instance, micro-interactions—small, subtle animations or sound effects that respond to user actions—can create a sense of delight and engagement.

Emotional design also extends to how software handles negative emotions, such as frustration or confusion. Interfaces that offer helpful, empathetic feedback when something goes wrong can mitigate user frustration and create a more positive overall experience. For example, a well-designed error message that offers a clear explanation and a solution can turn a frustrating situation into an opportunity for user trust.

Future trends in emotional design will focus on creating more personalized, emotionally engaging experiences that cater to users' psychological needs. This might include interfaces that adapt their tone and style based on the user's emotional state, creating a more human-like interaction.

Affective Computing: Emotion Recognition and Response

Affective computing takes emotional design a step further by enabling systems to detect and respond to users' emotions in real time. By analyzing facial expressions, voice tones, gestures, and physiological signals (such as heart rate or skin conductivity), affective computing systems can gauge how a user is feeling and adjust the interface accordingly.

For example, an educational app might detect when a student is becoming frustrated or disengaged and adjust the difficulty level or offer encouragement. Similarly, a fitness app could recognize when a user is tired or stressed and modify its workout recommendations to better suit the user's current state.

While effective computing has exciting potential, it also raises significant ethical and privacy concerns. Systems that can detect and respond to emotions must handle sensitive data responsibly, ensuring that users are aware of what information is being collected and how it is being used.

Usability in Autonomous Systems and Smart Devices

Autonomous systems, such as self-driving cars, drones, and smart home devices, are becoming increasingly prevalent, raising new questions about usability. These systems operate with a high degree of autonomy, making decisions and performing tasks without constant human input. As a result, the focus on usability in autonomous systems shifts from direct interaction to ensuring that users understand, trust, and can effectively manage these systems.

Usability in Autonomous Vehicles

Self-driving cars are one of the most visible examples of autonomous systems, and their usability is critical to their success. Users must trust that the vehicle can navigate safely and make sound decisions, but they also need clear feedback on what the vehicle is doing and why. For instance, autonomous vehicles must communicate their intentions—such as stopping, turning, or changing lanes—in a way that is easily understood by both the driver and other road users.

In addition to trust, autonomy introduces new challenges in terms of user control. Users must be able to take over control of the vehicle when necessary, and the system must provide clear, intuitive controls for doing so. Usability design for autonomous vehicles must strike a balance between offering the convenience of autonomy and maintaining user agency.

Usability in Smart Home Devices

Smart home devices, such as thermostats, lighting systems, and voice assistants, are becoming more autonomous, learning from user behavior and adjusting their settings automatically. While this can improve convenience, it also introduces new usability challenges. Users need clear feedback on what the system is doing and how decisions are being made, as well as the ability to override or adjust these decisions if desired.

For instance, a smart thermostat might learn a user's temperature preferences and adjust automatically, but the user must still have an easy way to change the settings when needed. Smart devices must be designed with intuitive interfaces that allow for both autonomous operation and manual control.

As more devices become interconnected through the Internet of Things (IoT), the usability of these systems will depend on creating seamless, cohesive experiences across multiple devices. Designers must ensure that users can easily manage and control complex ecosystems of smart devices, providing transparency and simplicity in interactions.

Ethical Considerations in Future Usability Trends

As software becomes more intelligent and pervasive, ethical considerations are becoming increasingly important in the design of usable systems. Trends in usability, particularly those involving AI, personalization, and affective computing, raise questions about privacy, data security, and the potential for manipulation.

Privacy and Data Security

Many of the emerging technologies discussed in this chapter—AI, affective computing, and personalized interfaces—rely on collecting large amounts of data about users, including their behaviors, preferences, and even emotions. Ensuring that this data is collected, stored, and used ethically is critical to maintaining user trust.

Designers must prioritize transparency, making it clear to users what data is being collected and how it will be used. Privacy settings should be easily accessible and understandable, allowing users to control what information they share. Additionally, data security measures, such as encryption and anonymization, must be implemented to protect sensitive user information from breaches or misuse.

Avoiding Manipulative Design

Personalization and AI-driven interfaces have the potential to shape user behavior in powerful ways, which raises ethical concerns about manipulation. For example, recommendation algorithms that are designed solely to maximize engagement or profit may lead users to spend more time on a platform than they intended or to make decisions that are not in their best interest.

Ethical design practices require that systems be designed with the user's well-being in mind. Rather than prioritizing profit or engagement metrics, designers should focus on creating interfaces that empower users, providing them with the information and tools they need to make informed decisions. Transparency is key—users should understand how algorithms influence their experience and have the ability to adjust or opt out of certain features if desired.

The Role of Bias in AI and Personalization

AI-driven systems are only as good as the data they are trained on, and if that data contains biases, the system's outputs will reflect those biases. For instance, AI algorithms used in hiring or loan approval processes have been shown to exhibit bias against certain demographic groups due to historical inequalities in the training data.

As AI plays a larger role in usability, it is crucial to address the issue of bias in algorithmic design. This requires diverse and representative training data, as well as ongoing monitoring and testing to ensure that AI systems do not perpetuate or amplify harmful biases. Ethical AI design should also involve mechanisms for accountability and transparency, allowing users to understand and challenge the decisions made by AI systems.

The Role of 5G and IoT in Shaping Usability

The rollout of 5G networks and the proliferation of the IoT will transform the way users interact with software, enabling faster, more connected, and responsive user experiences. These technologies will play a critical role in shaping the future of usability, particularly in terms of seamless cross-device interactions and real-time responsiveness.

5G's Impact on Usability

5G technology offers significantly faster data transfer speeds and lower latency compared to previous generations of mobile networks. This opens new possibilities for real-time, data-intensive applications such as augmented reality, virtual reality, and autonomous systems.

With 5G, users can expect faster load times, smoother streaming, and more responsive interactions, particularly in applications that rely on real-time data. For instance, AR applications can deliver more accurate and detailed information overlays without lag, while VR experiences can become more immersive and fluid.

The reduced latency of 5G also enhances the usability of interconnected devices in smart homes and cities. Users will be able to control and monitor IoT devices in real time, with instant feedback and updates. This creates opportunities for more dynamic and interactive user experiences, where devices respond to user actions with minimal delay.

IoT and Usability

The Internet of Things connects a wide range of devices, from household appliances to industrial machinery, creating a network of interconnected systems that can share data and automate tasks. IoT devices are becoming increasingly autonomous, using AI to learn from user behavior and adjust their operation accordingly.

Usability in IoT devices focuses on creating intuitive ways for users to interact with and manage complex ecosystems of connected devices. This includes providing clear feedback on device status, offering simple controls for managing multiple devices, and ensuring that users can easily understand how devices are communicating and making decisions.

One of the key challenges in IoT usability is ensuring that the system remains understandable and manageable as the number of connected devices grows. Designers must create interfaces that simplify the complexity of IoT systems, offering users clear insights into how their devices are interacting and providing easy ways to control or override automated actions.

Let's address potential limitations.

- **Security concerns**: Increased connectivity brings increased risks. Security vulnerabilities in IoT devices can compromise not just data but also physical safety (think connected cars or medical devices). Usability must consider security, with clear mechanisms for authentication, data encryption, and user control over access permissions.

- **Digital divide**: While 5G promises widespread benefits, its rollout might be uneven, creating a digital divide. Designers need to consider users with varying levels of connectivity and ensure their products are accessible and usable even in less-than-ideal network conditions.

- **Cognitive overload**: An abundance of connected devices can lead to information overload and user anxiety. Usability must focus on simplifying interactions, providing clear visual cues, and prioritizing information to avoid overwhelming users.

Accessibility and Inclusive Design

As software becomes more integrated into everyday life, ensuring that it is accessible to all users is a critical component of usability. Trends in usability must prioritize inclusive design, ensuring that users with disabilities or varying levels of technical expertise can interact with software without barriers.

AI-Powered Accessibility Features

AI and machine learning are playing an increasingly important role in improving accessibility. For users with visual impairments, AI-powered **screen readers** can provide real-time descriptions of on-screen content. **Speech-to-text technology** can assist users with hearing impairments by converting spoken language into written text.

Voice recognition systems, powered by AI, allow users with motor impairments to control devices through voice commands. **Image recognition** technologies help visually impaired users navigate their surroundings by identifying objects or reading signs.

These AI-driven accessibility features are continuously improving, offering users more natural and efficient ways to interact with software. As these technologies advance, we can expect to see even more sophisticated accessibility tools that cater to a wider range of needs.

Designing for Diversity

Inclusive design goes beyond adding assistive features—it involves considering the needs of diverse users from the very beginning of the design process. This means designing interfaces that are flexible, customizable, and usable by people with a wide range of abilities.

For example, offering adjustable text sizes, customizable color schemes, and alternative input methods (such as voice, touch, or gestures) ensures that software is accessible to users with visual, auditory, or motor impairments. Additionally, designing interfaces that are simple and intuitive can help users with cognitive impairments or those who are less familiar with technology.

In the future, usability innovations will continue to prioritize inclusivity, ensuring that as technology advances, it remains accessible to all users, regardless of their abilities.

Usability Testing with Emerging Technologies

As new technologies like AI, AR/VR, and IoT become more prevalent, usability testing methods must evolve to address the unique challenges these systems present. Traditional usability testing methods may not be sufficient for evaluating the complex, adaptive, and context-aware systems of the future.

Usability Testing in AI-Driven Systems

AI-driven systems, which adapt and evolve over time, require continuous usability testing. Unlike traditional software, where testing can be done once and the results remain valid, AI systems may change their behavior as they learn from user interactions. This means that usability testing must be an ongoing process, with periodic evaluations to ensure that the system remains usable and intuitive.

In addition to testing the system's usability, AI-driven systems must also be evaluated for transparency and explainability. Users need to understand how the system is making decisions, and testing should focus on whether the system's decision-making process is clear and understandable to users.

Testing AR/VR Interfaces

AR/VR usability testing presents unique challenges due to the immersive nature of these technologies. In AR/VR environments, users interact with virtual objects and spaces, requiring new methods of evaluation beyond traditional usability tests. For instance, testers must assess how intuitive it is for users to navigate 3D spaces, interact with virtual objects, and understand spatial relationships.

Motion sickness is another critical factor in VR usability, and testers must evaluate how well the system mitigates this issue. Additionally, AR systems must be tested for how well virtual objects integrate with the physical environment, ensuring that they are contextually relevant and easy to manipulate.

As these technologies continue to evolve, usability testing will need to become more sophisticated, incorporating new methods and tools to assess the effectiveness of next-generation interfaces.

Data-Driven Design: Analytics and Usability

In the future, data will play an increasingly important role in shaping software usability. Data-driven design involves using analytics and user behavior data to inform design decisions, allowing designers to create more efficient and user-friendly interfaces.

The Role of User Analytics in Usability Design

User analytics provide insights into how users interact with software, revealing patterns in behavior that can inform design improvements. For example, analytics can show which features users engage with most frequently, which areas of the interface cause confusion or frustration, and how long it takes users to complete specific tasks.

By analyzing this data, designers can identify pain points in the user experience and make targeted improvements. For instance, if analytics reveal that users are abandoning a process at a particular step, designers can investigate why and make changes to simplify that step or provide additional guidance.

Data-driven design also allows for continuous optimization of the user experience. By regularly analyzing user data, designers can track the impact of changes and make iterative improvements based on real-world usage.

Predictive Analytics for Personalized Experiences

In addition to understanding past behavior, predictive analytics can be used to anticipate future user needs and preferences. By analyzing patterns in user data, predictive algorithms can suggest personalized content, recommend features, or even adjust the interface to match the user's predicted actions.

For example, a predictive analytics system might notice that a user frequently accesses a particular tool at a certain time of day and proactively offers shortcuts or reminders related to that tool. This type of anticipatory design reduces friction and improves overall usability by offering relevant options before the user has to ask for them.

As predictive analytics become more advanced, they will enable even more personalized and adaptive user experiences, further enhancing usability.

Cognitive Load Reduction in Future Interfaces

One of the key challenges in usability design is managing cognitive load—the mental effort required to use a system. As interfaces become more complex, with multiple features and options, there is a risk of overwhelming users. Upcoming trends in usability may focus on reducing cognitive load by simplifying interactions and providing users with clear, concise information.

Streamlining Interfaces for Simplicity

A major trend in cognitive load reduction is the move toward minimalist design. By reducing the number of elements on the screen and focusing on the most essential features, designers can create interfaces that are easier to navigate and less mentally taxing.

For example, many modern mobile apps use flat design principles, with simple icons, limited color schemes, and minimal text. This type of design reduces visual clutter and helps users focus on the task at hand.

Another strategy for reducing cognitive load is progressive disclosure, where advanced features or options are hidden until they are needed. This allows users to start with a simple, easy-to-understand interface and gradually access more complex functionality as they become more familiar with the system.

Contextual Assistance and Guidance

Future interfaces will also incorporate more intelligent guidance systems to reduce cognitive load. These systems can provide real-time assistance, offering tips, suggestions, or explanations based on the user's current actions. For example, a design tool might offer contextual tooltips or shortcuts when the user hovers over a particular feature, helping them understand how to use it without needing to consult a manual.

AI-powered guidance systems can also adapt to the user's skill level, offering more detailed explanations for novice users and more advanced options for experienced users. This type of adaptive guidance helps reduce cognitive overload by ensuring that users only see the information they need at the moment.

The Role of Blockchain in Usability

Blockchain technology, known for its decentralized and secure nature, is primarily associated with cryptocurrencies and decentralized finance. However, its potential in software usability is gaining recognition, particularly in areas related to security, transparency, and trust.

Usability Benefits of Blockchain

Blockchain can enhance usability by providing more secure and transparent interactions between users and systems. For instance, decentralized identity management allows users to maintain control over their personal information, logging into multiple platforms without needing to create separate accounts for each service. This streamlines the user experience and reduces the cognitive load associated with managing multiple passwords and accounts.

In addition, blockchain can facilitate trust in transactions, as every action taken on a blockchain is recorded in an immutable ledger. This transparency is particularly valuable in areas like supply chain management, where users need to verify the authenticity of goods or trace their origin.

Challenges and Future Directions

Despite its potential, blockchain introduces new usability challenges, particularly in terms of complexity. Many blockchain-based systems require users to understand technical concepts such as wallets, private keys, and gas fees, which can be confusing for non-technical users.

For blockchain to become more usable, designers must focus on creating intuitive interfaces that hide the complexity of the underlying technology. Simplifying the process of managing blockchain assets and interacting with decentralized applications (dApps) will be critical to making blockchain technology accessible to mainstream users.

Conclusion

The future of software usability is being shaped by a range of emerging technologies, from AI and voice interfaces to AR/VR and blockchain. These innovations are redefining how users interact with software, creating more intuitive, adaptive, and personalized experiences.

As technology continues to evolve, usability design must keep pace, focusing on creating interfaces that are not only functional but also accessible, ethical, and emotionally engaging. Designers must consider the diverse needs of users, ensuring that software is inclusive and responsive to individual preferences, contexts, and abilities.

The trends explored in this chapter—such as AI-driven personalization, cross-platform consistency, emotional design, and data-driven optimization—are just the beginning of a new era in usability. As these technologies mature, they will continue to revolutionize how we interact with software, making it more seamless, intelligent, and human-centered.

In conclusion, the future of software usability is bright, with innovations that promise to make technology more accessible, efficient, and enjoyable for all users. However, as these trends progress, it is crucial to balance innovation with ethical considerations, ensuring that usability remains a force for good in enhancing the human experience.

CHAPTER 13

Driving Software Usability with Change Management

The Importance of Software Usability

The significance of software usability lies in its ability to shape the interaction between a user and an application, directly influencing the user's ability to accomplish their objectives. When software is highly usable, it is characterized by ease of learning and use, efficiency in performing tasks, and overall user satisfaction. These elements are critical as they significantly affect user productivity, satisfaction levels, and the software's market success. In today's world, where technology is deeply woven into the fabric of daily life, the usability of a software product is a key determinant of its acceptance and enduring relevance.

Usability is key because it helps users do what they need with software easily and happily. It's important because it makes people more productive and satisfied and helps the software do well in the market. Nowadays, when we use technology for almost everything, how easy software is to use can decide if it's a hit or a miss. Making software usable is not just about making it work; it's about making it work in a way that people like. So, when creators focus on making their software easy and enjoyable to use, they're not just improving a product. They're making technology friendlier for everyone. This doesn't just help the software sell better right now; it also makes sure it stays relevant as technology keeps changing.

Overview of Change Management

Change management helps us get ready for, adjust to, and support changes so that organizations can succeed. When we talk about making software, change management is super important. It makes sure that when we update or change how the software works, people can get used to these changes easily and like them. This process includes many steps, like making plans for the changes, checking them to make sure they work, telling people about them, and listening to what they think. All this effort is to make sure people are okay with the new changes and like using the updated software.

In making software better or adding new things to it, change management acts like a helpful guide. It starts before the change happens by figuring out what needs to be better and making a plan. Then, it helps by teaching and preparing everyone who will use the software, making the transition smoother. By talking clearly about what's changing and why and by listening to users' feedback, change management works to reduce any worries or problems people might have. This way, everyone can move forward together, embracing new features that make the software better for everyone's use.

Linking Usability and Change Management

Merging the concepts of usability and change management is about understanding that making software easier to use isn't just about tweaking the software itself. It's also about changing the way people use it. This idea means that if we want to make software better for users, we need to guide them through these changes carefully. This guidance comes through a planned approach to change management.

A well-thought-out change management plan is crucial when we introduce new features or make existing ones easier to use. This plan ensures that users are not just aware of the new changes but are also comfortable using them. It teaches them how to integrate these changes into their daily tasks, making the transition smooth. This approach reduces any negative reactions or resistance to the change, ultimately making the software more pleasant and efficient to use.

By taking users by the hand through each step of the change, addressing their concerns, and showing them the benefits of the new features, we can greatly improve their experience. This not only makes them more likely to embrace the new changes but also enhances their overall satisfaction with the software. In turn, this leads to more efficient work processes and a more positive interaction with the technology.

Understanding Software Usability

Definition and Key Components

Software usability delves into the simplicity and speed with which individuals can accomplish their desired tasks using a software application. This concept is broad, capturing various essential elements that contribute to an effective and user-friendly software experience.

375

- **Speed of task completion**: This element questions how swiftly users can complete their tasks within the software. It's about reducing the time taken to achieve objectives and enhancing productivity.

- **Task accuracy and completeness**: This focuses on the user's ability to execute their tasks accurately and to completion without errors. It measures the effectiveness of the software in helping users reach their goals precisely.

- **Ease of learning**: Here, the emphasis is on how straightforward it is for new users to understand and get accustomed to the software. This element is crucial for reducing the learning curve and enabling users to become proficient with the application more quickly.

- **Memorability**: This aspect evaluates whether users can easily remember how to use the software after a period of not using it. It's important for infrequent users who need to re-engage with the application without going through a relearning process.

- **User satisfaction**: This measures the overall comfort and positive feelings users have when using the software. User satisfaction is key to ensuring continued use and positive recommendations.

These essential elements collectively lay the groundwork for assessing and improving software usability. They serve as critical guidelines for developers and designers, pushing them to craft applications that are not just functional but also enjoyable and efficient to use. By prioritizing these components, software creators can ensure their products holistically meet the users' needs, fostering a satisfying and productive user experience.

Challenges to Software Usability

Making software easier to use comes with its own set of challenges.

- **Different users**: People using the software come from various backgrounds, have different levels of skill, and like different things. This makes it hard to create software that fits everyone perfectly.

- **More complex tasks**: As software gets more advanced, it can do more complex things. This is great, but it can also make the software harder for people to understand and use without getting confused.

- **Keeping it simple yet powerful**: It's tempting to add lots of features to make software do more. However, this can make the software more difficult to use. The key is to add enough features so the software is useful but keep it simple enough so people aren't overwhelmed.

- **Listening to users**: It's very important to hear what users say about the software and make changes based on their suggestions. But, gathering this feedback and making changes can take a lot of time and effort.

Each of these challenges requires careful thought and planning to overcome. The goal is to make software that's powerful enough to do what users need but easy enough for them to use without frustration. This involves a lot of testing, listening to user feedback, and making continuous improvements.

Strategies for Enhancing Usability

To tackle these hurdles, you can use a few smart strategies.

- **Design with users in mind.** This means including users in the development process from start to finish. By doing this, we make sure the software fits what users want and need. It's like asking for directions when you're not sure where you're going—it helps you get there faster and with fewer wrong turns.

- **Make things simpler.** By making the software's design cleaner and its steps more straightforward, we help users not to feel overwhelmed. Think of it as decluttering a room so you can find what you need without any hassle.

- **Test with real users often.** Checking in with users at different points while making the software lets us catch and fix any problems early. It's like tasting your food while you're cooking it so you can make sure it's coming out just right.

- **Think about everyone.** It's important to make sure people with disabilities can also use the software. This is not just the right thing to do, but it also means more people can use the software. Imagine making sure there's a ramp so everyone can come into your store— it's about including everyone.

- **Scaling with change management**: Using change management helps us introduce new features smoothly. It means getting users ready for updates, helping them adjust, and hearing their thoughts to make the software better. Imagine guiding someone through a new area they haven't seen before, making sure they're comfortable, and answering their questions. This approach makes changes easier for users, builds their trust, and keeps them happy.

With change management, as the software gets better and bigger, we make sure users are with us every step of the way. This makes expanding the software's features and user base more manageable and ensures users stay satisfied as the software evolves.

- **Boosting user adoption**: Ensuring users embrace new features is crucial. We must motivate them to try these updates, similar to teaching friends a new game. This involves straightforward explanations, guidance, and occasionally incentives. Making adoption easy and showing the benefits are key. By concentrating on user adoption, our improvements have a real impact, enhancing the software for all users.

By following these strategies, we can create software that's not just powerful but also easy and enjoyable for everyone to use. It's about paying attention to the small details that make a big difference in making software friendly for all users. This approach encourages ongoing improvement, ensuring that the software evolves in a way that continues to meet users' changing needs and expectations.

Basics of Change Management

What is Change Management?

Change management is a planned approach to help an organization update its goals, processes, or technology. Its main goal is to make these changes smoothly and fully, trying to minimize any problems while getting the most out of the change. In software development, change management is very important for adding new features or updates. It ensures that users can easily accept these new changes and that they don't interfere with their daily work.

Models of Change Management

There are several straightforward methods organizations can use to handle changes. The following are three popular methods.

- **ADKAR model**: This method focuses on helping each person through the change. It involves five steps: realizing why change is needed, wanting to make the change, knowing how to change, being able to make the change, and keeping the change in place. This approach looks at the change from each person's perspective.

- **Kotter's 8-step process**: This method is all about the bigger picture in an organization. It starts with making sure everyone feels that the change is important, getting the right team together, making a clear plan for the change, telling everyone about it, making sure everyone can help make the change happen, celebrating the small victories along the way, keeping the momentum going with more changes, and making sure the new way of doing things sticks around.

- **Lewin's change management model**: Lewin's approach sees change as three steps: getting ready for change, making the change, and making sure the change lasts. It's about preparing for the change, doing it, and then making it a normal part of how things are done.

These methods offer different ways to look at and manage change, from focusing on individual experiences to changing the whole organization's culture to help make transitions smoother and more successful.

Change Management in Software Development

In software development, change management plays a crucial role in ensuring that updates or new features are smoothly integrated into the user experience. In the world of creating software, managing change is key to making sure that new updates or features fit well with how people use the software. Managing change effectively involves the following:

- **Planning**: Figuring out what changes need to be made, like which features to update or add, and thinking about how these changes affect users.

- **Talking to users**: Telling users what changes are coming, why they're happening, and how they will make things better. Clear and regular talks can help ease worries and get people on board with the changes.

- **Helping users adjust**: Giving users the tools they need to get used to the new features or updates. This could be through guides, frequently asked questions and support services.

- **Listening to users**: Setting up a way to hear what users think about the changes and adjusting based on their input. This makes sure the changes are what users want and need.

Managing change in software isn't just about the technical side; it's also about looking after the people using the software, making sure they're prepared, happy, and able to use the new parts of the software.

Integrating Usability with Change Management

Planning for Usability Changes

Starting to make software easier to use with change management needs careful planning first. This step means finding out exactly what parts of the software could be better for users and how making these parts better fits with what the business wants to achieve and what users need. Here's how to plan these improvements.

- **Understanding users**: It's crucial to dive deep into what users do, what they need, and the problems they face with the software. This can include doing surveys, talking directly to users, and watching how they interact with the software through usability tests. The goal is to get a full picture of the user experience from the users' point of view.

- **Setting goals**: You need to clearly state what you want these improvements to accomplish. Goals should be clear and detailed, possible to measure, achievable, directly tied to what's important, and have a specific timeline. This SMART (specific, measurable, achievable, relevant, and time-bound) approach ensures that each goal is meaningful and reachable, and it helps everyone understand what success looks like.

- **Bringing everyone together**: It's important to get everyone who has a stake in the software involved from the beginning. This includes the people who build and design the software, the users, and the people in charge. Getting everyone on the same page from the start makes sure that the changes will be useful and that everyone supports them. It also opens a line of communication for sharing ideas and feedback, which can lead to even better solutions.

By focusing on these key areas, you lay a strong foundation for integrating usability improvements effectively. It ensures that every change is made with a clear purpose and is backed by a deep understanding of user needs and business goals. This approach not only leads to a better product but also aligns everyone's efforts toward a common objective, fostering a more collaborative and user-focused development process.

Communicating Usability Changes

Good communication is essential for managing change well. It's important to keep everyone who uses or is affected by the software in the loop about what changes are coming, why they're happening, and how they will make things better. The following describes ways to communicate effectively.

- **Straightforward messaging**: It's crucial to use simple, clear language to describe what's changing. Avoid technical jargon that might confuse people who aren't experts.

- **Using many ways to share information**: To make sure the message gets through to everyone, it's a good idea to use a variety of methods. This could include sending out emails, publishing updates in company newsletters, hosting online meetings, or having discussions in team meetings.

- **Staying in touch**: Keeping an ongoing conversation is key. This means not just announcing changes once and forgetting about it but regularly sharing updates. It also means giving people chances to ask questions or give their thoughts on the changes as they happen.

It's also helpful to tailor the communication to different groups. For example, what developers need to know about a change might be different from what end-users need to know. Similarly, setting up a dedicated space, like a forum or a Q&A session, where users can express concerns, suggest improvements, or simply learn more about the changes can enhance understanding and buy-in. Additionally, visual aids like diagrams or videos explaining the changes can be particularly effective for those who find visual information easier to understand than written explanations.

This approach to communication ensures that everyone affected by the changes feels considered and informed, which can significantly smooth the transition process. It helps build a sense of trust and collaboration, making it more likely that the changes will be accepted and adopted successfully.

Implementing and Reinforcing Changes

After planning and talking about the changes, it's time to actually make these changes happen. This step is all about turning the plans into action and making sure users get on board with the new updates.

- **Step-by-step introduction**: Bringing in changes bit by bit makes the transition easier to handle. It also means we can tweak things along the way based on what users tell us.

- **Help and learning**: Offering lessons, guides, and help to users so they can get used to the changes. This might include online guides, help centers, and instructions.

- **Watching and listening**: Keeping an eye on how things are going and asking for user opinions. This is crucial for spotting any problems early and fixing them.

- **Encouragement**: Praising and giving something extra to users who adjust well to the changes. This could be a simple thank you, rewards, or sharing stories of how the changes have helped.

Making these usability improvements work well with change management means having a plan that deals with both the tech side and the people side. It's about making sure everyone who's affected feels ready, included, and supported from start to finish.

Evaluating Usability Improvements
Methods for Measuring Usability

Checking if making software easier to use actually works is key. It helps us see what's working and what needs more work. There are several ways to do this.

- **Usability testing**: Watching people use the software to see where they get stuck or confused.

- **Surveys and questionnaires**: Asking users directly about their experience, how easy the software is to use, if they're happy with it, and any problems they've encountered.

- **Analytics**: Using software tools to track how people use the software, like which buttons they click most, how they move through the software, and where they run into errors. This helps spot where users are having trouble.

- **Heuristic evaluation**: Having experts look at the software using a checklist of good design practices to find usability issues.

These methods give a rounded view of how well the software meets users' needs and where it could be better, guiding improvements to make the software even easier to use.

User Feedback and Iteration

User feedback is essential for constantly making software easier to use. By creating a system for feedback, developers can update the software in small steps, using suggestions from actual users. The following are ways to do this.

- **Easy feedback tools**: Adding features in the software that make it simple for users to give their thoughts.

- **Scheduled feedback reviews**: Planning regular times to go over user feedback and decide which ideas to act on.

- **Choosing what to change first**: It's not possible to work on all feedback right away. Deciding what to change first should be based on how big an impact it will have, how often the problem comes up, and what it takes to make the change.

This approach helps make sure that the software keeps getting better in ways that matter to users.

Case Studies: Successes and Lessons Learned

Looking at examples of when making software easier to use worked well can teach us a lot. These examples should cover the following.

- **The problems**: What issues with ease of use did they find, and how did they figure them out?

- **The solutions**: What specific steps did they take to make the software easier to use, and how did they manage these changes?

- **The results**: What difference did these changes make to how happy and efficient users were and to how well the software worked?

- **What was learned**: What important lessons did they learn, and how can these help make software better in the future?

By measuring the effects of these changes, listening to what users say, and learning from what has worked (or hasn't) before, we can keep improving software in ways that help users.

Challenges and Solutions

Common Challenges at the Intersection of Usability and Change Management

Mixing in usability improvements with managing changes often runs into certain obstacles that can slow things down. The following are some common issues.

- **Not wanting to change**: People might not want to switch from what they're used to because they're comfortable, they're afraid of what's new, or they think it will make their work harder.

- **Not enough resources**: Not having enough time, money, or staff can limit how much you can improve usability and how well you can manage introducing these changes.

- **Different priorities**: When people involved in the project, like the developers, bosses, and users, don't agree on what's most important, it can cause disagreements about what changes should be made first.

- **Trouble communicating**: If the way we share information isn't working well, it can lead to confusion, people not supporting the changes, and missing out on valuable suggestions from others.

Expanding a bit more, overcoming these challenges requires creative solutions and persistence. For example, easing resistance to change might involve more personalized training sessions or demonstrations of the benefits of the new system to make the transition more appealing.

Addressing resource constraints could involve seeking additional funding sources or prioritizing the most impactful usability improvements within existing budgets. Bridging the gap between different priorities may require facilitated discussions or workshops to align everyone's goals. Improving communication might mean adopting new tools or methods that make sharing information and collecting feedback more effective and engaging for all parties involved.

By tackling these issues head-on, organizations can smooth the path for usability enhancements and change management, leading to a more user-friendly and efficient system.

Overcoming Resistance to Change

Overcoming resistance to change is key to making usability improvements work. Here are ways to do this.

- **Involve users early.** Getting users involved in the changes from the start helps them feel like they're part of the process and more likely to accept the changes.

- **Show the benefits.** Make it clear how the changes will make things better, like making tasks easier, improving satisfaction, and reducing mistakes.

- **Provide training.** Make sure users have the training they need to get comfortable with the changes. This helps ease worries and boosts their confidence.

- **Listen to users.** Give users a way to share their thoughts and ideas. This makes them feel heard and can offer useful feedback.

These approaches can help smooth the transition to new improvements, making users more willing to embrace change.

Ensuring Long-Term Usability and Adaptability

To make sure usability improvements last and keep getting better, we can use a few important strategies.

- **Keep improving.** Set up a system where you regularly collect feedback, check how things are going, and make necessary tweaks. This keeps the software easy to use and up-to-date with what users need.

- **Make flexible designs.** Create software that's easy to update or change. This way, you can make improvements later without causing problems.

- **Encourage creativity.** Support a work environment where people value learning from users, trying new things, and being open to changes. This helps the software grow and improve over time.

- **Watch for new ideas.** Keep an eye on new trends and technologies in software design. This helps you bring in fresh, useful features before they become standard.

Dealing with the challenges of making software easier to use and managing change well means tackling things from several angles. It's about getting people involved early, talking things through clearly, teaching users how to adapt, and setting up ways to keep making things better.

Future Directions
Emerging Trends in Software Usability

The way we use software is always changing, thanks to new technology and what users expect. Some major trends are shaping what comes next.

- **Voice and conversational interfaces**: With more devices listening and responding to voice commands, software is evolving to let people use their voice to interact, making technology easier and more natural to use.

- **Augmented reality (AR) and virtual reality (VR)**: AR and VR are making software more user-friendly by creating immersive ways to handle complex tasks or explore digital spaces in new ways.

- **Artificial intelligence (AI) and machine learning**: These technologies are making software smarter, allowing it to learn from how people use it to get better and easier to use over time.

- **Accessibility and inclusivity**: Making software that everyone can use, including people with disabilities, is becoming more important. This means creating technology that's open to everyone.

The Evolving Role of Change Management

As software becomes easier to use, managing change effectively is becoming even more essential. Looking ahead, the following are some key trends for managing change in this area.

- **Becoming more agile**: Change management needs to be flexible and quick, matching the fast pace of software updates.

- **Focusing on users**: Making sure changes are really about what users need and want, based on their feedback.

- **Helping with digital updates**: Assisting organizations as they adopt new technologies, making sure these changes are smooth and everyone is on board.

- **Encouraging flexibility and strength**: Helping organizations and people get better at dealing with change, seeing it as a chance to improve.

These trends highlight the importance of adapting and focusing on users to manage change successfully in the world of software.

Preparing for the Future of Usability Enhancement

Organizations and professionals should do the following to keep up with the fast-paced world of software usability.

- **Keep learning.** Always update your knowledge about usability and how to manage change by learning new things.

- **Work together.** Make sure people from different areas, like developers, UX designers, and users, work together to bring different views into the design.

- **Use data and feedback.** Rely on what data and user feedback tell you to make sure improvements really help users.

- **Try new things.** Build an environment where trying out new ideas is encouraged to find better ways to improve usability.

The future of making software easy to use is exciting, with new technologies and methods appearing all the time. By staying adaptable and ready for change, organizations can make sure their software stays top-notch for users.

Conclusion

Summary of Key Points

This chapter delved into the crucial blend of enhancing software usability and managing change, emphasizing their combined role in improving user experience and software adoption. The following are some key highlights.

- The significance of software usability in helping users achieve their goals with ease and satisfaction

- The function of change management in easing users and organizations into software updates, facilitating smooth transitions and acceptance of enhancements

- Strategies to align usability enhancements with change management, such as effective planning, communication, implementation, and reinforcement, aiming for seamless integration into user workflows

- Addressing challenges at the usability and change management intersection, like overcoming resistance, managing limited resources, and maintaining adaptability for future improvements

- Future trends in software usability and change management spotlight advances like voice interfaces, augmented reality, and AI, and there is a focus on greater accessibility and inclusivity.

Final Thoughts on Usability and Change Management

Combining software usability with managing change is key to creating software that's easy to use and keeps up with user needs. As technology advances quickly, being good at managing change and improving usability is essential.

Software development teams must focus on both usability and managing change. They should keep learning, work together well, and always think about what users need. This way, their software stays useful, current, and helpful.

To sum up, improving software usability and managing change is a continuous effort. It needs hard work, the ability to change, and a focus on users' needs. Looking ahead, we should be ready for both the chances and challenges of making software that makes a difference in users' lives.

CHAPTER 14

Challenges in Achieving Software Usability

Software usability, defined as the degree to which a software system is easy to use and meets users' needs, is a vital aspect of software success. Usability determines how effectively users can interact with the system to achieve their goals and how satisfied they are with the overall experience. However, designing software that achieves a high level of usability is fraught with challenges. These difficulties arise from the need to accommodate diverse user requirements, balance simplicity with functionality, and adapt to constantly evolving technology.

This chapter explores the complex landscape of software usability, examining the major challenges that developers and designers face in creating intuitive, efficient, and enjoyable user experiences. The discussion covers everything from understanding user requirements to addressing cognitive overload and managing the trade-offs between complexity and simplicity.

© Saurav Bhattacharya 2025
P. Gujar, *Data Usability in the Enterprise*, https://doi.org/10.1007/979-8-8688-1183-8_14

Usability is not just a superficial aspect of software design. It influences the fundamental structure of how software is built, evolves, and interacts with its users. Poor usability can result in user frustration, decreased productivity, and ultimately, the failure of a software product. Thus, achieving usability requires not only careful planning and design but also an ongoing commitment to monitoring and refining the user experience.

Defining Software Usability

Software usability encompasses several dimensions that collectively determine how easily users can accomplish their tasks within a system. These dimensions include learnability, efficiency, memorability, error handling, and user satisfaction. However, usability is more than just the sum of these parts—it also involves understanding user behavior, expectations, and the context in which the software is used.

At its core, software usability aims to minimize the effort required by users to complete tasks while maximizing their overall satisfaction with the experience. Usability can be evaluated through various means, including user testing, analytics, and feedback. However, achieving a high level of usability requires anticipating user needs and challenges long before the software is released. The design process must prioritize user-centered thinking from the outset, incorporating feedback loops and iterative improvements.

The Importance of Usability

The importance of usability in software design cannot be overstated. Whether developing consumer applications, enterprise software, or specialized tools for niche industries, the usability of the system plays a critical role in determining its success. Poor usability can render even the most feature-rich systems ineffective, as users struggle to accomplish basic tasks or grow frustrated with the interface.

Impact on User Adoption

Usability is directly tied to how quickly and widely users adopt a new software product. If users find the system difficult to navigate or understand, they are unlikely to continue using it. This is especially true in competitive markets where alternatives are readily available. Early impressions of a system's usability can make or break its success.

Enhancing Productivity

In business environments, usability has a direct impact on productivity. Systems that are easy to use and efficient allow employees to complete tasks more quickly and with fewer errors. This translates into significant time and cost savings for organizations. On the other hand, poorly designed software can lead to lost time, training costs, and frequent calls to support teams.

Reducing User Errors

Usability is also a key factor in reducing user errors. When systems are intuitive, users are less likely to make mistakes, and when they do make errors, the system should provide clear feedback and recovery options. Error-prone software frustrates users and can even lead to costly mistakes in high-stakes environments, such as healthcare or finance.

Building User Loyalty

A well-designed user interface (UI) fosters loyalty. When users enjoy using the software and feel confident in their ability to navigate it, they are more likely to continue using it and recommend it to others. This creates positive word-of-mouth marketing and encourages repeat usage, which is crucial for software-as-a-service (SaaS) models and consumer applications.

The Five Usability Principles

Software usability is generally understood through five core principles, each of which addresses a key aspect of the user experience. These principles serve as a guide for designers and developers in creating systems that are both functional and user-friendly. However, they also introduce significant challenges when trying to implement them in practice.

Learnability

Learnability refers to the ease with which new users can learn to use a system effectively. When software is intuitive, users can quickly pick up how to navigate its features without extensive training. Learnability is particularly important for applications aimed at the general public or infrequent users, such as mobile apps or online services.

Challenges

Designing for learnability often involves balancing ease of use with the need for more advanced features. While beginners need simple and clear instructions, experienced users might find too much guidance patronizing or limiting. This creates tension between simplifying the UI for new users and maintaining efficiency for more experienced users.

Efficiency

Efficiency refers to the speed and effort required for experienced users to complete tasks. Systems that support efficiency allow users to achieve their goals with minimal effort. This is particularly important in professional environments, where time savings translate into cost savings.

Challenges

Improving efficiency often requires incorporating shortcuts, customization options, and advanced features, which can complicate the interface for novice users. Striking the right balance between supporting fast task completion for experts and maintaining clarity for new users is a major challenge in usability design.

Memorability

Memorability is the extent to which users can remember how to use the system after a period of not using it. If users can easily recall the steps needed to perform tasks when they return to the system, it is more likely to be considered usable.

Challenges

Creating memorable interfaces requires consistent design patterns, clear navigation, and intuitive workflows. However, this can be difficult when dealing with complex software or systems that undergo frequent updates. Each update risks disrupting users' memory of how to interact with the system, especially if significant UI changes are made.

Error Prevention and Handling

Error prevention and handling focus on minimizing the likelihood of user mistakes and providing clear recovery paths when errors occur. Usable software should guide users away from making errors through intelligent design choices and, when mistakes happen, offer helpful feedback and solutions.

Challenges

While error prevention can be addressed through thoughtful design (e.g., disabling buttons until required fields are filled), error handling is often more complex. Designers must anticipate a wide variety of user errors and ensure that the system provides appropriate and meaningful guidance to help users recover from mistakes.

Satisfaction

Satisfaction refers to how enjoyable the system is to use. Beyond simply getting the job done, users should feel that the software is pleasant and satisfying. Satisfaction is subjective and can vary depending on user preferences, experience levels, and the context in which the software is used.

Challenges

Satisfying all users is impossible, given that different individuals have different expectations and levels of expertise. What one user finds satisfying (e.g., a minimalistic design) may be frustrating for another (e.g., an advanced user seeking more functionality). The challenge lies in striking a balance that maximizes satisfaction across the broadest range of users.

Challenges in Defining User Requirements

Defining user requirements is the foundation of successful software usability, but it is also one of the most difficult aspects of the design process. Understanding what users need and how they expect to interact with the software is crucial for developing a system that meets their expectations.

Implicit Needs

Users often have difficulty articulating their exact needs, particularly when it comes to usability. They may know what tasks they want to accomplish but not how they want to accomplish them. Additionally, users may not be aware of potential usability problems until they encounter them in practice.

For example, in developing a complex business application, users may request features without understanding the usability trade-offs involved, such as added complexity in the interface or slower performance due to processing overhead.

Changing Requirements

User needs and expectations evolve over time, influenced by new technologies, market trends, and changing business requirements. This means that what was considered usable at the start of a project may no longer be sufficient by the time the product is launched.

For example, a mobile app designed for basic e-commerce functionality might need to support new payment methods or integrate with other platforms (e.g., social media, cloud services), complicating the UI and potentially reducing usability if not managed properly.

Conflicting User Goals

Different users may have conflicting goals and expectations for how they interact with the software. Novice users typically prioritize ease of learning, while expert users may focus on efficiency and customization. Balancing these competing goals can be challenging.

For example, in productivity software, novice users may prefer simplified task lists and visual guides, whereas power users demand advanced shortcuts and customization options, such as drag-and-drop functionality or keyboard commands.

401

User Diversity

User diversity presents one of the most significant challenges in achieving software usability. A system must cater to a wide variety of users, each with different levels of experience, preferences, cultural backgrounds, and physical abilities. Designing for such diversity requires a deep understanding of the user base and a flexible approach to interface design.

Varying Skill Levels

The usability of a system must account for a wide spectrum of skill levels. Novices often need extensive guidance, tutorials, and clear workflows, while experts require efficiency, shortcuts, and customization options.

Challenges

Designing for multiple skill levels is difficult because what benefits one group may hinder another. An interface with step-by-step tutorials may frustrate experienced users, while complex, feature-rich interfaces can overwhelm beginners. Some systems address this by offering different modes (e.g., beginner, intermediate, expert) or by allowing users to toggle between simplified and advanced views.

Cultural and Language Differences

In globally used software, cultural differences can affect how users interpret certain design elements. Colors, symbols, metaphors, and even interaction patterns may carry different meanings in different cultures, making it challenging to design universally understandable software.

Challenges

Ensuring that software is usable across multiple languages and cultural contexts requires careful localization. Simple translation of text is often insufficient; cultural norms and expectations around navigation, interaction, and user feedback must also be taken into account.

Personal Preferences

Each user has unique preferences for how they interact with software. Some users may prefer visual interfaces with drag-and-drop functionality, while others favor keyboard shortcuts or voice commands. Accommodating these preferences while maintaining a consistent, coherent user experience can be difficult.

Challenges

Providing too many customization options can lead to complexity, cognitive overload, and maintenance challenges. Conversely, limiting personalization options may alienate users who feel that the software does not align with their preferred way of working.

Cognitive Overload

Cognitive overload occurs when a user is presented with too much information or too many choices at once, leading to confusion, frustration, and mistakes. This is a major challenge in usability design, particularly for complex systems with many features and functions.

Causes of Cognitive Overload

- **Complex interfaces**: Cluttered or poorly organized interfaces can overwhelm users, making it difficult for them to find what they need. This is especially problematic in software that tries to serve multiple purposes or includes many advanced features. For example, a dashboard displaying dozens of widgets, each with real-time data updates, may overwhelm a user who is only looking for specific information.

- **Information overload**: Providing users with too much information at once can result in decision paralysis. Users may struggle to determine the most relevant information, leading to slower task completion and increased errors. For example, a multi-step checkout process in an e-commerce app that asks for numerous optional details (e.g., gift wrapping, donation options, delivery instructions) can overwhelm the user, causing them to abandon the purchase.

- **Poorly organized navigation**: Disorganized or unclear navigation can confuse users and increase the cognitive load required to move through the software. Users should always know where they are in the system and how to reach their desired destination. For example, a website with multiple nested menus and inconsistent labeling of sections can make it difficult for users to find specific content.

Addressing Cognitive Overload

To reduce cognitive overload, designers must streamline the user experience and present information in a clear, hierarchical manner. Techniques for managing cognitive load include the following.

- **Progressive disclosure**: Only reveal the most relevant information at any given stage of the user's journey. Additional options or details can be shown when needed.

- **Minimalist design**: A clean, simple interface that prioritizes essential functions can help users focus on key tasks without being distracted by unnecessary elements.

- **Task-based navigation**: Break down complex tasks into smaller steps or phases to prevent overwhelming users with too many choices at once. For example, in a task-based productivity app, users can start by setting up basic tasks and then expand into more advanced features (such as project tracking and resource management) as they become more familiar with the software.

Balancing Functionality and Simplicity

Balancing the desire for rich functionality with the need for a simple, intuitive interface is one of the most difficult challenges in software usability. Complex systems often require advanced features, but these features can add complexity that reduces usability.

Feature Creep

Feature creep occurs when additional features are continuously added to a product, often in response to user requests or competitive pressures. While new features may enhance functionality, they can also clutter the interface, making it harder for users to navigate the system and accomplish core tasks.

Challenges

Feature creep often results from trying to satisfy too many users with different needs. As the system grows more complex, it can alienate both novice users (who are overwhelmed by the complexity) and expert users (who find it increasingly difficult to navigate to the tools they need).

Solution

A focus on core functionality, supported by user research and feedback, can help designers avoid feature creep. Prioritize features that align with the primary user goals and de-prioritize those that do not.

Designing for the Core User Experience

To maintain usability, software must focus on the most important tasks that users need to complete. This requires careful attention to the core user experience, ensuring that the most frequent tasks are easy to access and complete.

Challenges

Identifying the core user experience can be difficult, especially when designing for diverse user groups with varying needs. Additionally, there is a risk of oversimplifying the interface to the point where it no longer supports more advanced or specialized tasks.

Solution

User personas, user journeys, and task analysis can help identify the most important workflows, ensuring that they are prioritized in the design process. More complex or infrequent tasks can be made available but hidden from the primary interface to prevent clutter.

Providing Customization

Customization is often presented as a solution to the conflict between functionality and simplicity. By allowing users to personalize their experience, software can offer advanced features for power users while maintaining a clean interface for beginners.

Challenges

Customization introduces its own set of usability challenges. Allowing too much customization can confuse users, especially if the customization options are difficult to find or understand. Additionally, maintaining consistency across a customizable interface can be difficult.

Solution

Customization should be implemented carefully, with clear defaults and easy-to-use controls. Settings menus should be intuitive, and the consequences of customization options should be immediately visible to users.

For example, productivity software might allow users to customize their dashboard by adding or removing widgets. But the default dashboard should be simple and provide essential functions to ensure that all users can easily get started.

Usability in Complex Systems

Complex systems such as enterprise resource planning (ERP) systems, content management systems (CMS), or customer relationship management (CRM) tools present unique usability challenges. These systems often have hundreds or thousands of features designed to support a wide range of tasks, users, and workflows. Making such systems usable for all potential users is a daunting task.

Complexity vs. Usability

The primary challenge in complex systems is managing the trade-off between the need for rich functionality and the desire for a simple, intuitive user experience. While complex systems must support a wide range of tasks, users may find the interface overwhelming, leading to poor adoption and productivity.

For example, an ERP system used by a multinational corporation might include modules for finance, human resources, supply chain management, and sales. Each module has its own set of features, and users from different departments interact with the system in different ways.

Solution

Task-based workflows, user role-specific interfaces, and the use of progressive disclosure can help manage complexity. These techniques allow the system to display only the necessary information for a given user or task, reducing cognitive load and improving usability.

Workflow Integration

In complex systems, users often move between different modules or components of the system to complete tasks. Ensuring that these workflows are seamless and intuitive is critical to usability.

Challenges

Each module or component may be developed by a different team, resulting in inconsistencies in design or functionality. Users may become disoriented if workflows are not logically organized or if they need to switch contexts frequently.

Solution

Consistent design patterns, shared navigation structures, and clear visual hierarchies can help maintain a sense of continuity across different parts of the system. Additionally, ensuring that data flows smoothly between modules reduces the need for users to manually transfer information, improving workflow efficiency.

Training and Support

While usability should always be a priority, some complex systems will inevitably require user training, especially in professional environments where the software supports specialized tasks. In these cases, providing comprehensive training resources, including tutorials, documentation, and user guides, is essential.

Challenges

Relying too heavily on training can indicate underlying usability problems. If a system requires extensive training before users can perform basic tasks, it may need to be simplified. At the same time, users with advanced roles may still require training to understand specialized functionality.

Solution

Balancing training with good design is key. Provide basic training for new users but ensure that the system is intuitive enough that users can learn on the job without needing constant reference to documentation. For advanced users, targeted, role-specific training can help them leverage the full power of the system.

Accessibility Challenges

Accessibility ensures that software can be used by individuals with a wide range of disabilities, including visual, auditory, motor, and cognitive impairments. While accessibility is an important aspect of usability, achieving it presents several challenges.

Common Accessibility Issues

- **Visual impairments**: Users with visual impairments, including blindness and color blindness, may struggle with software that relies heavily on visual elements or does not provide text alternatives for images and icons. For example, a data visualization tool that uses color alone to distinguish between data sets may be unusable for color-blind users.

- **Motor disabilities**: People with motor disabilities may find it difficult to use interfaces that require precise input, such as small buttons or complex drag-and-drop interactions. For example, a form that requires users to click small checkboxes or buttons may be difficult to navigate for users with limited dexterity.

- **Cognitive disabilities**: Users with cognitive disabilities may have difficulty understanding complex interfaces, processing large amounts of information, or following long workflows with multiple steps. For example, a multi-step registration process that requires users to remember and input a series of details (e.g., personal information, security questions, payment information) without clear guidance may overwhelm users with cognitive impairments.

Accessibility Best Practices

To improve accessibility, software should be designed with the needs of disabled users in mind from the outset. Some best practices include the following.

- **Keyboard navigation**: Ensure that all functionality can be accessed using a keyboard. Users who cannot use a mouse should be able to navigate the software easily through keyboard shortcuts and tab orders.

- **Screen reader support**: Provide text alternatives for all non-text content, such as images, icons, and graphs. Ensure that the software is compatible with screen readers used by visually impaired users.

- **High-contrast mode**: Offer high-contrast color schemes to accommodate users with low vision or color blindness.

- **Adjustable font sizes**: Allow users to increase or decrease font sizes according to their preferences, ensuring that text remains readable for users with visual impairments.

411

- **Simple language:** Use plain, simple language to improve understanding for users with cognitive disabilities. Avoid technical jargon or complex sentence structures.

Challenges

Implementing these best practices can sometimes conflict with other usability goals, such as aesthetics or the desire for minimalism. However, accessibility should never be sacrificed for visual appeal. Additionally, ensuring compatibility with a wide range of assistive technologies can be time-consuming and requires thorough testing.

Cross-Platform Usability

In today's multi-device world, software is often expected to run seamlessly across multiple platforms, including desktop computers, mobile devices, and web browsers. Designing software that is usable on all these platforms presents its own challenges.

Device Constraints

Each platform comes with its own constraints and user expectations. Mobile devices, for example, have smaller screens, limited input methods (e.g., touch rather than a mouse and keyboard), and different interaction patterns. Desktop software, on the other hand, may support more complex interactions and larger data sets.

Challenges

Designing for multiple platforms requires a deep understanding of each platform's limitations and strengths. Interfaces that work well on one platform may not translate well to another. For example, a complex menu structure might work on a desktop but be unusable on a mobile phone's smaller screen.

Solution

Adopt a mobile-first approach to ensure that the core functionality of the software is usable on small screens. Then, gradually enhance the interface for larger devices, ensuring that the software feels natural on each platform.

Responsive Design

Responsive design involves creating software that adapts to different screen sizes and input methods. A responsive interface automatically adjusts its layout and functionality to provide an optimal user experience across devices.

Challenges

Designing truly responsive interfaces is not easy. Developers must account for various screen sizes, resolutions, and input methods while maintaining a consistent user experience. Inconsistent behaviors between platforms can confuse users, leading to frustration and reduced usability.

Solution

Use flexible grids, scalable images, and media queries to ensure that the layout adapts smoothly across devices. Focus on core tasks that are relevant across all platforms and avoid platform-specific designs that may break when accessed from different devices.

Consistency vs. Platform-Specific Design

While users expect consistency in functionality across platforms, they also expect software to follow platform-specific design conventions. For example, users might expect certain navigation patterns in mobile apps (e.g., swipe gestures) that differ from those in desktop applications.

Challenges

Balancing consistency with platform-specific design can be difficult. Designers must decide whether to prioritize consistency across platforms (which might reduce usability on some devices) or to create platform-specific designs that better align with user expectations on each platform.

Solution

Identify core elements that should remain consistent across platforms (e.g., branding, core functionality) while allowing for flexibility in interaction patterns. Mobile apps, for instance, might include swipe-based navigation, while the desktop version uses a traditional menu bar.

Usability Testing and Its Limitations

Usability testing is a crucial part of ensuring that software meets user needs and expectations. However, despite its importance, usability testing comes with some challenges and limitations.

Representative User Testing

One of the main challenges in usability testing is ensuring that the participants accurately represent the software's target audience. If the test participants are too experienced or unfamiliar with the system, the results may not reflect the usability issues that real users encounter.

For example, a usability test for a professional graphic design tool may yield skewed results if all the participants are novice designers. The feedback will not reflect the needs of the tool's core user base—experienced professionals.

Solution

Recruit test participants who closely match the target audience in terms of experience, needs, and context of use. If the software is intended for a broad user base, include participants from different segments (e.g., beginners, intermediates, and experts).

Resource Constraints

Usability testing requires time, personnel, and equipment, all of which may be in short supply, particularly in smaller development teams or startups. Additionally, running comprehensive usability tests on multiple platforms can be resource-intensive.

Challenges

Limited resources may force teams to skip or rush usability testing, leading to usability issues that go unnoticed until after launch. This can result in costly redesigns or reduced user adoption.

Solution

Consider using remote usability testing tools, which allow users to test the software in their own environment. This can reduce costs and increase the sample size. Alternatively, guerrilla testing methods can be used, where quick, informal usability tests are conducted with real users in public spaces.

Test Bias

Usability tests can be biased if users feel pressured to provide positive feedback or if they are aware of the presence of the development team. This bias can result in misleading test results, where usability issues are overlooked.

For example, in a lab-based usability test, users may feel uncomfortable criticizing the software if they know the development team is observing them, leading to skewed results.

Solution

Conduct tests in a neutral environment and assure participants that their honest feedback is valuable, regardless of whether it is positive or negative. Avoid having the development team directly interact with the participants during the test to reduce bias.

Measuring Subjective Experiences

Usability testing can objectively measure factors such as task completion time or error rates, but capturing subjective experiences like user satisfaction is more difficult. Surveys and interviews provide insight into how users feel about the software, but interpreting this feedback can be challenging.

Challenges

User satisfaction is highly subjective and may vary widely between individuals. Moreover, satisfaction scores alone do not always provide actionable insights into how to improve the user experience.

Solution

Combine quantitative and qualitative methods in usability testing. For example, track metrics like task completion rates and error frequencies while also conducting post-test interviews to gather subjective feedback. Analyzing the relationship between objective performance metrics and subjective satisfaction scores can offer deeper insights into the software's usability.

Continuous Usability Improvement

Usability is not a static goal that can be achieved once and forgotten. It requires continuous attention throughout the software's lifecycle, especially as user needs evolve, new features are added, and the software is used in new contexts.

Post-Launch Monitoring

Once a product is launched, it is essential to monitor how users interact with the software in real-world conditions. Usability problems that did not surface during development or testing may become apparent as more users interact with the system. Tools such as analytics, heatmaps, and user session recordings can provide valuable insights into user behavior.

Challenges

Post-launch monitoring can reveal a flood of usability issues, but it can be difficult to determine which issues are critical and require immediate attention. Additionally, fixing usability issues after launch can be costly, especially if significant redesigns are needed.

Solution

Use analytics tools to track key metrics, such as task completion rates, bounce rates, and user engagement. Prioritize usability issues based on their frequency and impact on the overall user experience. Regularly release updates to address critical issues and improve usability.

Iterative Design

Usability improvements should be iterative, with new versions of the software being released to address issues and refine the user experience. This iterative approach aligns well with agile development methodologies, where feedback loops and regular releases are integral to the process.

Challenges

Iterative design requires careful management of user expectations. Frequent updates may confuse or frustrate users, especially if significant changes are made to the interface. There is also a risk of introducing new usability problems with each update.

Solution

Incorporate user feedback into the design process and prioritize usability improvements based on their impact. Clearly communicate the purpose of updates to users, particularly when major changes are made. Provide training resources or tutorials to help users adjust to new features or workflows.

Agile Usability Practices

Agile development methodologies emphasize flexibility, collaboration, and rapid iteration, all of which support continuous usability improvement. However, integrating usability practices into agile workflows can be challenging.

Challenges

Agile development teams often prioritize speed and feature development, which can leave little time for thorough usability testing or refinement. Additionally, usability specialists may find it difficult to keep up with the rapid pace of development sprints.

Solution

Embed usability specialists within agile teams to ensure that usability is considered throughout the development process. Incorporate regular usability testing into sprints, even if it is done on a smaller scale. Use rapid prototyping and feedback loops to iterate on designs and refine usability before final implementation.

Usability and User Feedback Management

User feedback is a valuable source of information for improving software usability, but effectively managing this feedback is a challenge. Users often have conflicting opinions, and not all feedback is actionable or relevant.

Categorizing Feedback

Feedback should be categorized based on its relevance to usability issues, feature requests, or performance concerns. Not all feedback directly relates to usability, so it is important to differentiate between requests for new features and genuine usability problems.

Challenges

Sorting through large volumes of feedback can be time-consuming, and it may be difficult to identify which issues are most pressing. Additionally, users may provide vague or contradictory feedback, making it challenging to understand their specific usability concerns.

419

Solution

Use tools that allow users to vote on feedback or categorize it by type (e.g., usability issue, feature request, bug report). This can help prioritize the most common and impactful issues. Consider conducting follow-up interviews or surveys to clarify vague feedback.

Prioritizing Usability Improvements

Not all usability issues can be addressed immediately. Improvements should be prioritized based on the severity of the issue, its frequency, and its impact on the overall user experience.

Challenges

Determining the relative importance of usability issues can be difficult, particularly when there are resource constraints or conflicting user opinions. Some issues may be easy to fix but have a low impact, while others may require significant redesigns but dramatically improve the user experience.

Solution

Develop a prioritization framework that considers factors such as issue severity, user impact, and development effort. Focus on fixing critical usability issues that affect the core user experience and plan larger usability improvements as part of longer-term development cycles.

Balancing User and Business Needs

There is often a tension between addressing user feedback and meeting business goals. For example, users may request simplifications to a feature that the business considers essential for driving revenue or differentiating the product from competitors.

Challenges

Addressing user feedback may conflict with business requirements, leading to difficult trade-offs. For example, simplifying a feature might improve usability but reduce its competitiveness or functionality, impacting business objectives.

Solution

Involve stakeholders from both the business and design teams in the prioritization process to ensure that usability improvements align with broader business goals. Consider the long-term impact of usability improvements on user retention, customer satisfaction, and brand loyalty.

Managing Expectations

When gathering and addressing user feedback, it is important to manage expectations. Not all feedback will result in immediate changes, and some requests may never be implemented due to technical or business constraints.

Challenges

Users may become frustrated if they feel that their feedback is being ignored, particularly if they have raised the same usability issue multiple times. Additionally, sudden or unexpected changes to the interface based on feedback can confuse or alienate users.

Solution

Communicate openly with users about how feedback is being addressed and what changes they can expect in future updates. Be transparent about the limitations and constraints that may prevent certain usability improvements from being implemented. Providing a public roadmap can help users understand which feedback is being considered and when they can expect changes.

Conclusion

Achieving software usability is a complex, ongoing process that requires balancing multiple factors, including user needs, business objectives, and technical constraints. While the principles of usability—such as learnability, efficiency, and satisfaction—provide a framework for design, the practical implementation of these principles is fraught with challenges.

Designing for diverse users, managing cognitive load, balancing functionality and simplicity, and addressing accessibility all require careful thought and iteration. Usability testing, feedback management, and post-launch monitoring further complicate the process as user expectations evolve and new issues arise.

Ultimately, achieving and maintaining high usability requires a commitment to continuous improvement, with a focus on the user experience throughout the software's lifecycle. By addressing the challenges outlined in this chapter, developers and designers can create software that not only meets functional requirements but also provides an intuitive, satisfying, and efficient user experience.

PART V

Future Trends

CHAPTER 15

The Road Ahead for Software Usability

Software usability, a field that once focused solely on making digital systems easy to use, has expanded dramatically in scope and influence over the past several decades. Usability is no longer confined to simplifying interfaces—it now encompasses how users emotionally, cognitively, and physically interact with software across diverse devices and environments. As the digital world grows more complex, with emerging technologies such as artificial intelligence, augmented reality, and voice-based interfaces gaining traction, the usability challenges and opportunities continue to evolve.

This chapter explores the road ahead for software usability, examining how foundational principles remain relevant in the face of changing technologies and shifting user expectations. It covers established usability concepts, emerging trends, and the new role that AI and multimodal interaction are playing in shaping user interfaces. The expansion of usability beyond traditional interfaces—touchscreens, keyboards, and mice—into realms like gesture-based, conversational, and immersive virtual environments demonstrates the importance of rethinking our approach to usability for the future.

© Saurav Bhattacharya 2025

P. Gujar, *Data Usability in the Enterprise*, https://doi.org/10.1007/979-8-8688-1183-8_15

With the growing importance of the user's experience, software developers must not only design systems that function efficiently but also create experiences that are intuitive, engaging, and emotionally satisfying. This chapter aims to provide a comprehensive view of usability's future, focusing on evolving trends, potential challenges, and the principles that will guide the development of more human-centered, inclusive software in the coming years.

Understanding Usability

At the most basic level, **usability** is about how effectively, efficiently, and satisfactorily users can accomplish tasks within a system. However, modern usability extends beyond ease of use; it touches upon every aspect of user interaction, including how software meets the user's cognitive needs, emotional states, and contextual realities.

ISO 9241-11: The Standard Definition

According to the ISO 9241-11 standard, usability is defined by three main attributes.

- **Effectiveness**: How accurately users can complete tasks using the system.

- **Efficiency**: How quickly and with what effort tasks are completed.

- **Satisfaction**: The comfort and overall perception of the system's use.

These factors combine to form the baseline for evaluating usability, but they are just the beginning of what it means to create truly user-friendly software. Usability is a discipline that intersects with user experience, encompassing emotional responses to software, as well as how intuitive, accessible, and contextually appropriate software feels for users.

Usability Beyond Interfaces

Usability considerations now include the following.

- **Cognitive load**: How much mental effort is required for users to interact with the system? Intuitive interfaces aim to reduce cognitive load by aligning the design with user expectations and behaviors.

- **Context of use**: Usability depends on where and how users engage with software. For instance, mobile applications used in busy environments require different usability considerations than desktop applications used in controlled office settings.

- **Task appropriateness**: Effective usability ensures that the software fits the specific goals and tasks of users, avoiding unnecessary complexity.

The concept of usability has expanded to reflect the sophisticated ways that users interact with devices, applications, and systems in today's fast-evolving technological landscape.

The Evolution of Software Usability

Software usability has evolved from the early days of human-computer interaction (HCI), when the primary concern was reducing the complexity of text-based command interfaces. The trajectory of usability's evolution mirrors technological advancements, particularly as systems transitioned from specialized technical tools to everyday consumer products.

Early Usability and the Birth of HCI

In the 1960s and 1970s, human-computer interaction was primarily the domain of engineers and computer scientists, who focused on making computer systems functional. However, these systems were difficult for non-expert users to understand. The introduction of graphical user interfaces (GUIs) in the 1980s marked the beginning of a significant shift, making computers more accessible by replacing complex command-line inputs with visually intuitive controls, like windows, icons, and menus.

The GUI Revolution: Making Technology Accessible

The advent of GUIs in operating systems such as Apple's Macintosh and Microsoft Windows revolutionized usability. GUIs provide a way for users to interact with software using visual metaphors that are closely aligned with the real world (e.g., "folders" for file storage, "trash bins" for deleted items). This period also saw the rise of usability testing as a key element of software development, driven by the need to ensure that non-technical users could interact effectively with computers.

Web and Mobile Usability

The Internet and the rise of web-based applications in the late 1990s and early 2000s brought a new set of challenges and opportunities for usability. **Web usability** focuses on making websites accessible to a broad audience with varying technical abilities. With the growth of e-commerce, usability became directly tied to business outcomes, as poor usability was linked to lower sales and user engagement.

Mobile usability presented even more challenges, as designers had to rethink interaction models for smaller screens and touch-based navigation. Mobile interfaces demanded simplicity, clarity, and responsiveness, pushing usability considerations to the forefront of design.

Usability in the Age of AI and Voice Interfaces

Today, usability continues to evolve with the proliferation of AI-powered systems, voice interfaces, augmented reality, virtual reality, and other advanced technologies. These emerging technologies require new usability strategies that move beyond traditional graphical interfaces, leading to more immersive and interactive user experiences.

The increasing integration of usability into business models—through concepts like user-centered design and user experience design—has positioned usability as a strategic differentiator in the success of software products.

Key Principles of Usability

Several core principles underpin effective software usability, guiding designers and developers toward creating interfaces that are efficient, intuitive, and satisfying.

Consistency

Consistency is a fundamental aspect of usability. It refers to maintaining uniformity in the design elements, layout, behavior, and interaction patterns throughout the interface. When users encounter familiar elements or interactions across different parts of the system, they can quickly form mental models that help them navigate the software intuitively. Consistency reduces cognitive load, allowing users to focus on their tasks rather than learning new interaction paradigms.

The following describes consistency types.

- **Visual consistency**: Refers to maintaining the same visual elements (such as colors, fonts, and icons) across different pages or screens within the software.

- **Functional consistency**: Involves ensuring that similar actions produce similar outcomes throughout the interface (e.g., the same keyboard shortcuts or gestures trigger the same functions).

- **External consistency**: Refers to consistency with industry standards and user expectations. For example, the placement of a "home" button on a website should follow established web design conventions.

Feedback

Effective feedback informs users about the results of their actions, helping them understand whether a task has been completed successfully or if further steps are required. Feedback can take many forms.

- **Visual feedback**: A common example is the change in color or appearance of a button after it has been clicked.

- **Auditory feedback**: Notifications or beeps can alert users to completed tasks, errors, or new information.

- **Tactile feedback**: On touchscreens, haptic feedback (such as vibrations) can confirm interactions like button presses or swipes.

Without feedback, users may feel disconnected from the system or unsure of what to do next, leading to frustration and inefficiency.

Affordance

Affordances are cues that suggest how an object or interface element can be used. For example, a button that appears raised or shaded implies that it can be clicked, while a text box with a blinking cursor suggests it is ready for input. Good affordances eliminate ambiguity and guide users toward correct actions without the need for detailed instructions.

In digital interfaces, affordances play a critical role in helping users understand the function of various elements. When affordances are intuitive, users can perform tasks more easily, and the learning curve is reduced.

Error Prevention and Recovery

Users make mistakes, but well-designed software can minimize errors and help users quickly recover when they occur. There are two key aspects to this principle.

- **Error prevention**: Involves designing systems that guide users away from making mistakes in the first place. Examples include form validation, constrained inputs (e.g., drop-down menus instead of free-text fields), and clearly labeled buttons.

- **Error recovery**: When errors do occur, the system should provide clear, actionable feedback to help users correct their mistakes. Descriptive error messages should explain what went wrong and how to fix it, while "undo" options can provide a quick way to reverse unintended actions.

Learnability

Learnability refers to how easily new users can grasp the interface and start using the system effectively. A high degree of learnability is crucial, particularly for complex applications, as it lowers the entry barrier for users and reduces the need for extensive training.

Best Practices for Improving Learnability

- **Progressive disclosure**: Rather than overwhelming users with all available features at once, display only the most essential tools and options at first, gradually revealing more advanced features as users become more proficient.

- **Contextual help**: Incorporate inline help features, tooltips, or brief tutorials to assist users when they encounter unfamiliar functions.

Learnability is especially important in software where first impressions can determine whether users will continue using the product or abandon it in favor of simpler alternatives.

Summary

In summary, software usability hinges on consistency, clear feedback, intuitive affordances, error mitigation, and learnability. These principles ensure user interfaces are efficient, enjoyable, and accessible to all.

User-Centered Design

User-centered design (UCD) is an iterative design approach that places the user's needs, behaviors, and goals at the forefront of the design process. By focusing on the end-user from the beginning, UCD ensures that the software is aligned with user expectations, preferences, and limitations.

Stages of User-Centered Design

The UCD process typically involves several stages, each aimed at refining the product based on user feedback and observations.

- **User research**: The foundation of UCD is understanding the target audience. User research may involve interviews, surveys, contextual inquiries, and personas to capture the needs, motivations, and pain points of users.

- **Prototyping**: Designers create **prototypes**, or simplified models of the software, which may include low-fidelity wireframes or interactive mockups. These prototypes allow for quick experimentation and validation of ideas before extensive development resources are committed.

- **Usability testing**: Real users are brought in to interact with the prototype, performing specific tasks while being observed. Usability testing provides insight into how well the design meets user needs and where improvements are necessary.

- **Iteration**: Based on feedback from usability tests, the design is iterated and refined. This process continues until the product meets usability and functional goals.

Benefits of User-Centered Design

- **Better alignment with user needs**: By focusing on user input, UCD helps create products that solve real problems rather than just providing technical solutions.

- **Reduced development costs**: Identifying and addressing usability issues early reduces the cost and effort associated with redesigning or fixing problems after launch.

- **Increased user satisfaction**: When users feel that software is designed with their needs in mind, their satisfaction and loyalty increase, resulting in higher adoption rates and retention.

Incorporating UCD principles ensures that usability is a continuous consideration throughout the development process rather than an afterthought.

Heuristic Evaluation and Usability Testing

To ensure that software meets usability standards, designers and developers use two main evaluation methods: heuristic evaluation and usability testing. Each has its strengths, and both can be used in conjunction to improve software usability.

Heuristic Evaluation

Heuristic evaluation is a usability inspection method where expert evaluators review the interface and compare it against a set of established usability principles, known as **heuristics**. This approach helps identify usability issues early in the design process before real users are involved. Common usability heuristics, as defined by Jakob Nielsen, include the following.

- **Visibility of system status**: The system should always keep users informed about what is happening through timely feedback.

- **Match between system and real world**: The system should speak the user's language, using familiar words, phrases, and concepts.

- **User control and freedom**: Users often perform actions by mistake. The system should provide an easy way to undo and redo actions.

- **Consistency and standards**: Users should not have to wonder whether different words, actions, or situations mean the same thing.

Heuristic evaluation is a quick and cost-effective way to uncover usability problems, but it relies on the expertise of the evaluators. As a result, it is often complemented by usability testing with real users to provide deeper insights.

Usability Testing

Usability testing involves observing real users as they interact with the software, allowing designers to identify how users behave, where they struggle, and what improvements can be made. Testing can be conducted at different stages of development, from early prototypes to post-launch versions.

Types of Usability Testing

- **In-person testing**: A facilitator observes the user performing tasks in a controlled environment. Users are often asked to "think aloud," explaining their thought process as they navigate the system. This approach allows designers to understand users' mental models and identify any disconnects between the interface and user expectations.

- **Remote testing**: Remote usability testing enables users to interact with the system in their own environment, providing insights into how the software is used in real-world settings. Remote testing tools record user interactions, clicks, and behaviors for later analysis.

- **Automated usability testing**: Tools like click-tracking software or heatmaps provide insights into where users are focusing their attention on the screen. These methods can be useful for testing at scale, providing data-driven insights into user behaviors.

Benefits of Usability Testing

- **Direct user feedback**: Usability testing provides direct insight into how users interact with the product, allowing developers to identify pain points and areas for improvement.

- **Actionable insights**: By observing users in real time, designers can uncover specific usability issues that may not be evident from heuristic evaluation alone.

- **Improved user satisfaction**: Continuous usability testing ensures that the final product is aligned with user expectations, reducing frustration and increasing overall satisfaction.

Accessibility in Usability

Accessibility is the practice of designing software that is usable by people with a wide range of abilities and disabilities. Accessibility is an essential component of usability, ensuring that software is inclusive and equitable for all users, regardless of physical, sensory, or cognitive impairments.

The Importance of Inclusive Design

Inclusive design focuses on creating products that are usable by as many people as possible without the need for adaptation or specialized design. By considering the needs of users with disabilities from the outset, designers can create interfaces that work for everyone. Inclusive design is especially important as digital technology becomes more pervasive in everyday life, from education and healthcare to commerce and entertainment.

Web Content Accessibility Guidelines

The Web Content Accessibility Guidelines (WCAG) provide a framework for ensuring that web content is accessible to all users. The guidelines are built around four key principles.

- **Perceivability**: Information and user interface components must be presented in ways that users can perceive, regardless of their sensory abilities. For example, providing text alternatives for non-text content (like images or videos) allows users with visual impairments to understand the content.

- **Operability:** The interface must be navigable by all users, including those who use assistive technologies like screen readers or who rely on keyboard-only navigation. Common practices include ensuring that interactive elements are accessible via keyboard shortcuts and that touch targets are large enough for users with mobility impairments.

- **Understandability**: The interface should be intuitive and easy to understand for all users, with clear language, instructions, and error messages. Content should be structured to help users predict the next steps and avoid confusion.

- **Robustness:** The software should be compatible with a wide range of devices and assistive technologies, ensuring that it can be used effectively across different platforms and environments.

Accessibility Tools and Techniques

Designers can leverage a variety of tools and techniques to ensure that software meets accessibility standards.

- **Screen readers:** Assistive technology that converts text on the screen into speech, allowing visually impaired users to interact with digital content. Designers must ensure that all interactive elements are properly labeled and that the screen reader conveys necessary information in a logical order.

- **High contrast mode:** High contrast modes help users with low vision or color blindness by improving the visibility of text, icons, and interactive elements.

- **Voice control:** Voice-based input systems, such as those used in mobile devices or smart speakers, provide an alternative means of interaction for users with mobility impairments.

Designing for accessibility not only benefits users with disabilities but also improves the overall usability for all users by ensuring that the software is flexible, adaptable, and user-friendly.

Usability and User Experience

While usability focuses primarily on how easy and efficient it is for users to accomplish tasks, the user experience encompasses the broader emotional, aesthetic, and psychological interactions that users have with a product. Usability is a critical aspect of the user experience (UX), which design goes beyond functionality to consider how the product makes the user feel and how well it meets their overall needs.

The Relationship Between Usability and UX

Usability is concerned with making systems intuitive, simple, and efficient, while UX addresses the holistic experience that users have with a product. A product with good usability allows users to accomplish tasks effectively, but a good UX design ensures that users enjoy the process and feel satisfied with the outcome. For example, a mobile banking app may be highly usable in terms of its functionality, but if the visual design is clunky or if users feel overwhelmed by too many notifications, the overall user experience may still be negative.

Emotional Design

One emerging area in UX design is emotional design, which emphasizes the creation of products that evoke positive emotional responses from users. Donald Norman, a cognitive scientist and pioneer in the field of UX, described emotional design as considering the user's feelings and reactions during their interaction with a product. The goal is to make products that are not only usable but also enjoyable and even delightful to use.

The following are elements of emotional design.

- **Aesthetics**: A visually pleasing interface can elicit positive emotions, making users feel more engaged and confident in using the product.

- **Micro-interactions**: Subtle animations, sound effects, or transitions can provide a sense of pleasure during interaction, reinforcing user satisfaction. For example, a slight bounce effect when a user completes an action can add a layer of enjoyment to the interaction.

- **Gamification**: By incorporating elements of game design (such as rewards, progress tracking, and challenges), designers can make tasks more engaging and rewarding, leading to higher user satisfaction.

Usability in Mobile and Web Applications

The shift to mobile and web-based applications has introduced new challenges and opportunities for usability. Designers must now account for smaller screen sizes, touch-based interactions, and variable network conditions, all while maintaining a seamless and consistent user experience across devices.

Mobile Usability

Mobile usability demands a streamlined approach to interface design, as users often interact with mobile devices in short bursts, while on the go, or in less-than-ideal conditions. Key considerations for mobile usability include the following.

- **Responsive design**: Mobile applications must be responsive, meaning that the interface automatically adjusts to fit different screen sizes and orientations, providing a consistent experience across smartphones, tablets, and other devices.

- **Touch-friendly interactions**: Designers need to optimize for touch gestures, such as swiping, pinching, and tapping. Touch targets should be large enough to avoid mis-clicks, especially for users with limited dexterity.

- **Performance**: Slow load times or unresponsive interfaces can severely impact the usability of mobile applications. Optimizing for speed is critical, especially in areas with poor network connectivity.

Web Usability

Web usability focuses on creating seamless, intuitive experiences for users navigating websites or web-based applications. As web applications become more complex, with features rivaling those of desktop software, usability remains a critical component of user satisfaction.

- **Navigation simplicity**: Web applications should have clear, straightforward navigation paths, reducing the number of clicks or interactions required to complete a task. Breadcrumbs, sticky headers, and search functionality are useful tools for improving navigation.

- **Cross-browser compatibility**: Ensuring that web applications perform consistently across different browsers (such as Chrome, Firefox, Safari, and Edge) is essential to maintaining usability. Inconsistent behavior across platforms can frustrate users and lead to higher abandonment rates.

Progressive Web Apps and Usability

Progressive web apps (PWAs) represent a hybrid approach to web and mobile usability, combining the best of both worlds. PWAs behave like mobile apps but are built using web technologies. They can be accessed directly through a web browser or installed on a user's device, blurring the line between traditional mobile apps and web experiences.

The following describes the advantages of PWAs in usability.

- **Offline access**: PWAs can function offline or in low-connectivity environments, ensuring that users can continue to access essential features even when their Internet connection is unstable.

- **Speed and responsiveness**: PWAs are typically faster and more responsive than traditional web applications, thanks to optimized caching and the use of service workers.

- **Consistency across devices**: PWAs provide a consistent experience across different platforms and devices, ensuring that users have a familiar interaction pattern whether they are on a mobile phone, tablet, or desktop.

By leveraging the unique features of PWAs, designers can create seamless and intuitive experiences that enhance usability while reducing development overhead.

Emerging Trends in Usability

As technology advances, new trends are emerging that push the boundaries of traditional usability design. The rise of voice interfaces, augmented reality, virtual reality, and wearable technology has introduced new paradigms for interaction, requiring fresh usability considerations.

Voice and Conversational Interfaces

Voice user interfaces (VUIs) allow users to interact with software through spoken commands. VUIs have grown in popularity with the rise of smart speakers (like Amazon Echo and Google Home) and voice assistants (like Siri and Alexa). While voice interfaces offer a hands-free, convenient way to interact with systems, they present unique challenges.

- **Understanding natural language**: VUIs must account for a wide variety of accents, speech patterns, and vocabularies. Ensuring that the system accurately understands and responds to users is key to maintaining usability.

- **Feedback in voice interfaces**: Since voice interfaces lack the visual cues of traditional GUIs, it is important to provide clear, verbal feedback to users about the status of their actions (e.g., confirming commands, providing progress updates).

Designing for voice interfaces involves a different mindset from GUI design, as it relies on conversational patterns and the human ability to communicate through language.

Augmented Reality and Virtual Reality

Augmented reality (AR) and virtual reality (VR) are immersive technologies that blend the physical and digital worlds, offering new possibilities for user interaction. However, they also introduce new usability challenges.

- **Spatial interactions**: Users must be able to navigate and interact with digital elements in a 3D space. This requires clear visual cues, intuitive controls, and thoughtful placement of virtual objects to avoid disorientation or confusion.

- **Reducing fatigue**: In VR environments, prolonged use can lead to physical discomfort, such as motion sickness or eye strain. Usability in these systems must prioritize user comfort, ensuring that interactions are simple, efficient, and not overly demanding on the user's physical or cognitive resources.

Designing for AR/VR systems requires a deep understanding of human perception, motor skills, and spatial reasoning, making usability even more crucial in these emerging fields.

Wearable Technology and Usability

Wearable devices, such as smartwatches, fitness trackers, and AR glasses, are becoming increasingly popular. Usability considerations for wearable devices focus on the following.

- **Context-aware interfaces**: Wearables are often used in dynamic, real-world environments, such as while exercising or commuting. Usability in this context means ensuring that information is displayed clearly and interactions are simple, even when the user is in motion or distracted.

- **Minimalist design**: Given the limited screen real estate on wearable devices, designers must focus on presenting only the most critical information and interactions, minimizing clutter and complexity.

Wearable devices also present opportunities for more contextual and personalized interactions, as they can access real-time data about the user's environment, location, and activity levels.

Artificial Intelligence and Usability

Artificial intelligence (AI) is reshaping the landscape of usability by enabling more personalized, adaptive, and predictive user experiences. AI-driven systems can analyze user behavior, predict needs, and automate tasks, all of which contribute to improved usability.

Personalization

AI can be used to deliver highly personalized user experiences by learning from user behavior and preferences. This can lead to interfaces that dynamically adapt to each user's individual needs.

- **Tailored content**: AI can recommend content or actions based on a user's previous interactions with the system, making the experience more relevant and reducing the time users spend searching for information.

- **Adaptive interfaces**: AI can modify the layout or functionality of an interface based on the user's behavior. For example, frequently used features can be highlighted or made more accessible, while less relevant features can be hidden.

Automation and Task Prediction

AI-driven automation can enhance usability by simplifying complex workflows and reducing the cognitive load on users. **Predictive algorithms** can anticipate user actions and offer suggestions or automate routine tasks.

- **Intelligent assistance:** AI can act as an assistant, automating tasks such as filling out forms, organizing data, or suggesting the next step in a process. This reduces the number of manual steps users need to perform and increases efficiency.

- **Contextual awareness:** AI systems can leverage contextual data (such as location, time of day, or previous actions) to proactively offer relevant information or recommendations. For example, a calendar app might automatically suggest scheduling changes based on travel delays or other disruptions.

Natural Language Processing

Natural language processing (NLP) enables software systems to understand and respond to human language. NLP is a key component of chatbots, voice assistants, and other conversational interfaces, which are becoming more common in customer service, e-commerce, and productivity applications.

NLP enhances usability by making interactions more intuitive and human-like. Users can communicate with the system using natural speech or text rather than having to learn specific commands or navigation paths.

AI and Usability Testing

AI is also transforming the way usability testing is conducted. AI-powered tools can automate aspects of usability testing, such as the following.

- **Analyzing user behavior:** AI can track and analyze user interactions, identifying patterns and detecting potential usability issues automatically.

- **Simulating user tests**: AI can simulate different user scenarios, allowing designers to test how the software responds to a variety of inputs and behaviors.

AI-driven usability testing provides faster feedback and deeper insights into user behavior, enabling designers to make data-driven decisions that improve the overall user experience.

Designing for a Multimodal Future

As technology advances, **multimodal interaction**—where users can interact with systems using a combination of voice, touch, gestures, and even gaze or motion—is becoming more common. Multimodal systems provide users with more flexibility and control over how they engage with software, allowing them to choose the input method that best suits their needs and context.

The Rise of Multimodal Interfaces

Multimodal interfaces offer several advantages for usability, particularly in situations where a single mode of interaction (such as touch or voice) may be limiting. For example, a user interacting with a smart home device might combine voice commands with touch gestures on a smartphone app to control their environment. Similarly, in a car navigation system, a user might speak an address aloud but tap the screen to confirm the destination.

Designing for Multimodal Usability

Designing for multimodal systems requires careful consideration of how different input methods interact with one another. It is important to ensure the following.

447

- **Input modes complement each other**: Different input modes should work together to enhance the user experience. For example, if a user starts an action using a voice command, they should be able to complete it using touch without any confusion or interruption.

- **Seamless transitions between modes**: Users should be able to switch between input modes effortlessly. The system should recognize when a user switches from one mode to another and adjust the interface accordingly.

Accessibility in Multimodal Interfaces

Multimodal systems also offer significant accessibility benefits, as they allow users with different abilities to choose the input mode that works best for them. For example, users with mobility impairments might prefer voice commands over touch, while users with speech impairments might rely on touch or gesture-based input.

By designing systems that accommodate multiple input modes, designers can create more inclusive and adaptable experiences that work for a diverse range of users.

The Psychology of Usability

Usability is deeply tied to human psychology because it deals with how users perceive, process, and respond to information. Understanding the cognitive processes that underlie user behavior can help designers create more intuitive and satisfying experiences.

Cognitive Load

Cognitive load refers to the amount of mental effort required to use a system. High cognitive load can lead to frustration, errors, and abandonment, while low cognitive load makes the system feel intuitive and easy to use. Designers can reduce cognitive load by doing the following.

- **Simplifying navigation**: Providing clear paths to complete tasks and reducing unnecessary complexity.

- **Chunking information**: Breaking down large amounts of information into smaller, more digestible pieces.

- **Using familiar patterns**: Leveraging familiar design patterns and interactions to help users form mental models of how the system works.

Attention and Focus

Humans have limited attention, and designing interfaces that capture and maintain focus is crucial for usability. Visual hierarchy, color contrast, and spatial organization can all be used to direct users' attention to the most important elements on the screen, helping them complete tasks more efficiently.

Error Aversion and Trust

Users are naturally averse to making errors, and a poorly designed system that leads to frequent mistakes can erode trust. To build trust and reduce anxiety, designers should do the following.

- **Provide clear feedback**: Let users know when actions are successful or when errors occur, and offer guidance on how to recover from mistakes.

- **Use predictable interactions**: Avoid surprises in how the system behaves. Predictability helps users feel more in control of their interactions.

Understanding the psychological principles that govern human behavior allows designers to create systems that are engaging and satisfying to use.

Usability and Cultural Contexts

Usability is not a one-size-fits-all concept; it can vary significantly depending on cultural contexts. Software that works well in one region or demographic group may need adjustments to be effective in another. Cultural differences can affect how users perceive, interpret, and interact with interfaces.

Cultural Differences in Design Preferences

Cultural preferences can influence everything from color choices and iconography to language and interaction styles. For example, users in some cultures may prefer minimalist interfaces with clean, simple designs, while users in other cultures may favor more vibrant, information-dense layouts.

Localization and Usability

Localization goes beyond simply translating text into different languages. It involves adapting the entire user experience to fit the cultural and linguistic expectations of the target audience. Key considerations for localization include the following.

- **Right-to-left (RTL) layouts**: Some languages, such as Arabic and Hebrew, are read from right to left. Ensuring that the interface supports RTL layouts is critical for usability in these regions.

- **Date and time formats**: Different regions use different formats for displaying dates and times (e.g., MM/DD/YYYY vs. DD/MM/YYYY). Designers must account for these variations to avoid confusion.

- **Cultural icons and symbols**: Certain icons or symbols may have different meanings in different cultures. For example, a "thumbs up" icon, which is generally positive in Western cultures, may be considered offensive in others.

By considering cultural contexts, designers can ensure that their software is usable and appealing to a global audience.

Usability for Advanced Technologies

As technology continues to evolve, designers face new challenges in making advanced systems—such as artificial intelligence, blockchain, quantum computing, and Internet of Things devices—usable and accessible.

Usability in AI Systems

AI introduces complexity to the user experience, as users must interact with systems that learn, adapt, and make decisions autonomously. Designers must ensure that users understand how AI systems work and how to interact with them. The following are key challenges.

- **Transparency**: Users need to understand how AI decisions are made. Designers should provide clear explanations of AI processes, especially in critical domains like healthcare or finance.

- **Control**: While AI systems can automate tasks, users must feel that they retain control over the system. Offering manual overrides or customization options can improve trust and usability.

Usability in Blockchain Systems

Blockchain technology, known for its decentralized and secure nature, presents usability challenges due to its complexity. The following are key considerations.

- **Simplifying interfaces**: Blockchain systems often require users to interact with complex concepts like wallets, keys, and ledgers. Simplified interfaces that abstract away the technical details can make blockchain more accessible to everyday users.

- **Security and privacy**: Blockchain's security features (such as the use of private keys) must be clearly explained to users, ensuring they understand how to manage their assets securely.

Usability in IoT Devices

Internet of Things (IoT) devices—such as smart home systems, wearables, and connected appliances—require new approaches to usability. The following are key challenges.

- **Device interoperability**: Users expect IoT devices from different manufacturers to work together seamlessly. Ensuring interoperability and a consistent user experience across devices is crucial.

- **Context awareness**: IoT devices are often used in dynamic environments where users' needs and contexts change rapidly. Designing adaptive, context-aware interfaces can improve usability by providing relevant information and controls when needed.

Challenges in Usability Design

Despite advancements in usability practices and tools, several challenges remain, especially as technology becomes more complex and users' expectations continue to rise.

Complexity and Feature Creep

As software systems become more feature-rich, maintaining simplicity and usability can become difficult. Feature creep—the gradual addition of new features over time—can result in bloated interfaces that confuse users and degrade the overall experience.

Balancing Innovation with Usability

Innovative technologies, such as augmented reality, gesture-based interactions, or brain-computer interfaces, offer new possibilities for user interaction. However, these innovations can also introduce usability challenges if users are not familiar with the new interaction models. Designers must strike a balance between leveraging cutting-edge technology and ensuring that the interface remains intuitive and familiar to users.

Designing for Diverse Audiences

As digital products reach a global audience, designers face the challenge of creating interfaces that work well for diverse users with different cultural, linguistic, and accessibility needs. Ensuring that software is inclusive and adaptable across various regions and user groups requires careful planning and continuous iteration.

The Future of Usability Metrics and Measurement

As usability continues to evolve, so too must the ways in which it is measured. Traditional metrics like task completion time, error rates, and user satisfaction remain important, but new metrics are emerging that capture the broader aspects of user experience.

Emotional Usability Metrics

Future usability assessments will likely include metrics that track users' emotional responses to software, such as frustration, delight, or engagement. Tools that analyze facial expressions, voice tone, or physiological signals (e.g., heart rate or galvanic skin response) could provide insights into the emotional side of usability.

AI-Powered Usability Testing

AI can assist in usability testing by automating the collection and analysis of user data. AI systems can simulate user interactions, detect potential usability issues, and even offer design recommendations based on user behavior patterns. AI-driven usability testing provides faster feedback, enabling designers to iterate more rapidly.

Conclusion

As we look to the future of software usability, several trends and challenges will shape the landscape. Emerging technologies—from AI and voice interfaces to AR/VR and multimodal interactions—are pushing the boundaries of how users interact with digital systems. The integration of usability with emotional design, accessibility, and globalization will ensure that software products are not only functional but also engaging, inclusive, and culturally appropriate.

Usability is no longer just about simplifying interfaces; it's about creating systems that seamlessly blend into users' lives, adapting to their needs, behaviors, and environments. The road ahead for usability is rich with opportunities to innovate but also fraught with challenges that will require designers to rethink traditional paradigms and continuously refine their approaches.

By staying adaptable, embracing new trends, and applying the core principles of usability, developers and designers can build software that enhances the quality of life for all users, making the digital world a more accessible, intuitive, and enjoyable place.

PART VI

Conclusion

CHAPTER 16

Conclusion

As we reflect on the evolution of software usability, it's clear that what began as a focus on simplifying user interfaces has transformed into a multifaceted discipline. Today, usability is integral to crafting intuitive, accessible, and emotionally resonant digital experiences. Far beyond mere functionality, modern usability encompasses the human side of technology—how people feel, interact, and engage with software. It factors in accessibility for all users, emotional responses, cultural diversity, and more, all within a rapidly evolving technological landscape.

This concluding chapter synthesizes the concepts explored throughout the book, offering a comprehensive perspective on the future of usability. It examines the critical role of usability in contemporary software design, the application of timeless usability principles, and how emerging technologies such as artificial intelligence (AI), voice interfaces, and multimodal interactions are shaping new horizons. Furthermore, it discusses the growing importance of collaboration, the business implications of good usability, and the continuing dynamic evolution of the discipline. Ultimately, usability will remain a driving force for innovation and human-centered design in the coming decades.

P. Gujar, *Data Usability in the Enterprise*, https://doi.org/10.1007/979-8-8688-1183-8_16

Usability's Central Role in Modern Software Design

In today's hyper-connected digital age, usability is no longer a peripheral consideration in software development; it is at the core of successful product design. As digital systems proliferate, usability is fundamental to making technology accessible, functional, and engaging. Whether the software is for entertainment, productivity, education, or health, its success hinges on how easy it is to use, how efficiently it allows users to achieve their goals, and how satisfying the experience feels.

From Utility to Delight: The User Experience Imperative

In earlier decades, usability was often confined to functionality—did the system allow users to complete their tasks? While this remains essential, usability now also encompasses the broader user experience (UX), considering how software makes users feel and whether the experience is delightful and engaging. User-centered design places users at the heart of the development process, ensuring that their needs, frustrations, and feedback are addressed from the earliest design stages through to product release. Good usability transforms basic interactions into seamless experiences that are a pleasure to use.

As the complexity of digital ecosystems grows, so too does the role of usability in shaping how users perceive value in software. Usability creates a competitive edge, as companies with intuitive, accessible products tend to see higher user retention and loyalty. As user expectations evolve, systems must deliver more than functionality—they must provide intuitive, integrated experiences that align with the cognitive and emotional needs of users.

Key Principles That Stand the Test of Time

Despite the rapid advancement of technology, certain usability principles remain steadfast in their importance. These principles, such as consistency, feedback, affordance, error prevention, and learnability, have guided the development of software for decades. However, their application continues to evolve in response to new interaction paradigms, emerging technologies, and diverse user behaviors.

The Continuing Relevance of Core Usability Principles

- **Consistency** ensures that interfaces behave predictably, helping users build mental models that allow them to navigate systems with ease. Regardless of how advanced technology becomes, the need for a consistent user interface remains critical for usability. Whether it's navigating a virtual reality environment or using a voice assistant, users rely on familiar interaction patterns to guide them through new digital landscapes.

- **Feedback** gives users real-time responses to their actions, helping them understand whether their actions were successful or if additional steps are required. As technology becomes more complex, feedback needs to be clear and immediate. In AI-driven systems, for instance, users may not understand the underlying logic behind recommendations or actions without proper feedback mechanisms in place.

461

- **Affordances** provide visual cues about how to interact with elements in an interface. In an era where touchscreens, gestures, and voice commands are becoming more common, affordances must be adapted to these new modalities while remaining intuitive for users. For example, users should understand how to interact with a virtual object in an augmented reality interface just as easily as they would with a physical object.

- **Error prevention and recovery** are increasingly important as systems grow in complexity. Preventing errors through well-designed workflows and intuitive controls while allowing for easy recovery when mistakes are made is crucial for user satisfaction.

- **Learnability** becomes a greater challenge as technology diversifies. Ensuring that users can quickly understand and use software, even when encountering entirely new interaction paradigms like virtual reality or AI, will continue to be a vital focus for usability designers.

Balancing Innovation with Familiarity

The challenge for usability designers is to balance the introduction of innovative features with the need for familiar interaction patterns. As technologies like voice interfaces, AI-driven systems, and mixed reality experiences become more widespread, the core principles of usability will help anchor users in new and sometimes unfamiliar environments. By maintaining a focus on consistency, feedback, and affordance, designers can help users navigate the rapidly changing technological landscape without feeling overwhelmed or alienated.

User-Centered Design: The Gold Standard for Usability

User-centered design (UCD) has long been the gold standard for usability because it places the user at the center of the design and development process. By continually engaging users throughout the design cycle—via research, prototyping, testing, and iteration—UCD ensures that the final product aligns with user expectations and behaviors.

The Iterative Nature of UCD

One of the strengths of UCD is its iterative nature. Each design phase is informed by user feedback, ensuring that usability issues are identified early and resolved before they become entrenched in the product. This iterative approach has proven especially useful as software systems become more complex and user needs more diverse. By continuously refining and improving based on real-world use, UCD allows for more flexible, user-centered solutions.

Scaling UCD for Emerging Technologies

While UCD has been effective in traditional software development, it must evolve to meet the demands of emerging technologies such as AI, virtual reality (VR), and augmented reality (AR). These technologies present unique challenges that require new methods for gathering user feedback and refining design. For example, AI-driven systems often involve decisions made by algorithms, which users may not fully understand. Ensuring the usability of these systems requires designing interfaces that can explain AI processes clearly and in an accessible way.

Similarly, designing for VR and AR environments involves creating intuitive spatial interfaces that leverage UCD principles in new contexts. Users interacting in a 3D space require different kinds of feedback, error prevention, and affordances compared to traditional 2D screen-based interfaces. UCD, with its iterative and user-driven focus, will continue to adapt to these new contexts, ensuring that even the most advanced systems remain usable and intuitive.

Emerging Technologies and Their Impact on Usability

The future of usability is being shaped by a range of emerging technologies that offer new possibilities for user interaction but also introduce significant challenges. These technologies, including artificial intelligence, multimodal interfaces, augmented reality, virtual reality, and wearable devices, are fundamentally changing how users engage with digital systems.

Usability in AI-Driven Systems

AI has the potential to revolutionize usability by making systems more adaptive and personalized. AI-driven systems can learn from user behavior, anticipate needs, and provide proactive assistance. However, designing usable AI systems requires careful consideration of transparency, control, and user trust. Users need to understand how AI makes decisions, especially in high-stakes applications like healthcare or finance. Ensuring that AI interfaces are explainable and offer clear feedback is critical to maintaining user confidence.

Multimodal Interfaces: Expanding Interaction Possibilities

Multimodal interfaces allow users to interact with systems using a combination of input methods, such as voice, touch, gestures, and gaze. This flexibility can greatly enhance usability, allowing users to choose the interaction mode that best suits their context and preferences. For example, in a smart home environment, users may issue voice commands while completing certain tasks via a touch interface on their smartphone.

However, multimodal systems also present new usability challenges. Designers must ensure that the various input methods work together seamlessly without causing confusion or cognitive overload. For instance, transitioning between voice commands and touch gestures should feel intuitive and fluid, with each mode complementing the other rather than introducing friction.

AR, VR, and Wearables: Immersive Usability

The rise of augmented reality, virtual reality, and wearable devices brings exciting possibilities for immersive experiences but also necessitates new usability considerations. In AR and VR environments, users often navigate in a 3D space, which introduces challenges around spatial awareness, interaction accuracy, and reducing user fatigue.

Wearable devices such as smartwatches, fitness trackers, and AR glasses present additional usability challenges related to screen size, context of use, and input methods. Designers must ensure that interfaces are optimized for small screens and that users can interact efficiently, even in dynamic, real-world environments.

As these technologies continue to develop, usability will play a critical role in determining how successfully they are adopted and integrated into everyday life. The goal will be to create seamless, intuitive experiences that leverage the strengths of each technology while minimizing complexity.

Artificial Intelligence: Transforming Usability Practices

AI is reshaping usability in profound ways, enabling systems to learn from user behavior, adapt to changing preferences, and automate routine tasks. By incorporating AI, software can become more responsive and personalized, reducing cognitive load and increasing efficiency for users.

AI-Driven Personalization and Adaptation

AI systems excel at personalizing user experiences by analyzing vast amounts of data and identifying patterns in user behavior. This allows AI to adapt interfaces dynamically, presenting users with the most relevant information or recommendations based on their previous interactions. For example, an AI-powered shopping app might learn a user's preferences over time, offering personalized product recommendations or streamlining the checkout process based on frequent purchase patterns.

This level of personalization enhances usability by reducing the amount of effort required to complete tasks and by making the system feel more intuitive. However, it also introduces new challenges around user control and transparency. Users may not always understand why certain recommendations are made, which can lead to frustration or distrust if the AI seems to behave unpredictably. Designing usable AI systems requires making the inner workings of AI transparent, offering users clear explanations for recommendations, and allowing them to override automated decisions when necessary.

Automating Usability Testing with AI

AI is also transforming how usability testing is conducted. Traditional usability testing involves observing real users as they interact with the system, but this can be time-consuming and resource-intensive.

AI-powered usability testing tools can automate aspects of the testing process, analyzing user behavior in real time and identifying areas where users struggle or experience friction.

By tracking user interactions, such as mouse movements, click patterns, or time spent on certain tasks, AI tools can provide valuable insights into the user experience. These insights can help designers quickly identify usability issues and prioritize areas for improvement, leading to faster iteration cycles and more efficient design processes.

The Challenge of AI Usability: Balancing Automation with Control

One of the key challenges of designing AI-driven systems is striking the right balance between automation and user control. While AI can automate many tasks, giving users more control over their interactions, it's important to ensure that users feel empowered, not overwhelmed. Usability designers must find ways to provide users with manual overrides and clear explanations of AI actions, ensuring that they retain a sense of ownership and control over their digital experiences.

Usability in the Age of Multimodal Interactions

Multimodal interfaces are changing how users interact with digital systems by allowing for multiple input methods—such as voice, touch, gestures, and gaze—often in combination. These interfaces are increasingly common in smart homes, connected vehicles, and wearable devices, offering users more flexibility in how they engage with technology.

Designing for Multimodal Flexibility

One of the biggest advantages of multimodal interfaces is the flexibility they provide. Users can choose the input method that best suits their context, such as issuing voice commands while cooking or using gestures to navigate a VR environment. However, this flexibility also presents challenges for usability designers.

Multimodal systems should ensure that the various input methods work together seamlessly. Users should be able to switch between voice, touch, and gestures without disruption or confusion. For instance, a user might begin a task by issuing a voice command and then complete it by tapping an on-screen button. The system should recognize this shift in input method and provide consistent feedback throughout the interaction.

Ensuring Usability Across Modalities

While multimodal interfaces offer greater flexibility, they also require careful consideration of each modality's strengths and limitations. For example, voice commands are ideal for hands-free interaction, but they may be less effective in noisy environments or when precise input is needed. Conversely, touch and gesture controls are more precise but may not be suitable for all contexts, such as when a user's hands are occupied.

Usability designers must ensure that each modality is optimized for the context in which it is used while providing clear affordances and feedback that guide users through the interaction. The goal is to create a unified experience that allows users to transition between input methods effortlessly without feeling disoriented or confused.

The Human Side of Usability: Psychology and Emotions

Usability is not just about functionality—it's also about how users think, feel, and respond to the software they use. Understanding the psychological and emotional aspects of usability is critical for creating interfaces that are not only efficient but also engaging and enjoyable.

Cognitive Load: Simplifying User Interactions

One of the primary goals of usability design is to reduce cognitive load, or the amount of mental effort required to use a system. When the cognitive load is too high, users may become frustrated, make mistakes, or abandon the system altogether. Designers can do the following to minimize cognitive load.

- Simplify navigation by providing clear, straightforward paths for users to follow.

- Break down complex tasks into smaller, more manageable steps.

- Use familiar interaction patterns to help users build mental models of the system.

Reducing cognitive load is especially important in systems that involve complex decision-making, such as financial software or healthcare applications. By presenting information clearly and avoiding unnecessary complexity, designers can help users feel more confident and in control.

Emotional Design: Creating Positive Experiences

In addition to cognitive factors, emotions play a significant role in how users perceive and interact with software. Emotional design aims to create interfaces that evoke positive feelings, such as trust, satisfaction, and delight. This can be achieved through the following.

469

- **Aesthetics**: Visually appealing interfaces that use color, typography, and layout to create a pleasing experience.

- **Micro-interactions**: Small, subtle animations or sound effects that add a sense of delight and engagement to routine tasks.

- **Personalization**: Tailoring the interface to individual users' preferences and needs, making the experience feel more personal and relevant.

By focusing on both cognitive and emotional factors, usability designers can create systems that not only work well but also foster a deeper connection between users and software.

Accessibility: Usability for All

As digital technology becomes more embedded in everyday life, ensuring that software is accessible to all users—regardless of physical, sensory, or cognitive abilities—is more important than ever. Accessibility is a key aspect of usability, and it requires designers to create systems that the widest possible audience can use.

Designing for Inclusivity

Inclusive design focuses on creating products that are usable by as many people as possible without the need for adaptation or specialized tools. This approach goes beyond simply meeting legal accessibility standards; it involves designing with the diverse needs of users in mind from the outset.

For example, designing a website with clear, high-contrast text and easily navigable menus not only benefits users with visual impairments but also improves usability for everyone, particularly in low-light conditions or on mobile devices.

Leveraging Assistive Technologies

Assistive technologies, such as screen readers, voice recognition systems, and alternative input devices, are critical for making digital content accessible to users with disabilities. Usability designers must ensure that their interfaces are compatible with these technologies and that they provide the necessary affordances for users to interact with the system effectively.

For example, ensuring that all interactive elements are properly labeled and that the system's content is logically structured can make a significant difference for users relying on screen readers. Similarly, designing touch interfaces with large, easy-to-tap buttons can help users with mobility impairments interact more easily.

The Global Reach of Usability: Cultural and Regional Considerations

As digital products reach a global audience, usability must take into account cultural, linguistic, and regional differences that affect how users perceive and interact with technology.

The Importance of Localization

Localization involves adapting software to meet the cultural and linguistic preferences of users in different regions. This goes beyond simply translating text; it includes adapting visual elements, navigation patterns, and even date and time formats to align with local norms and expectations. Crucially, localization should also account for varying literacy levels and include symbols or images that convey meaning without relying solely on text.

For example, a website designed for a Middle Eastern audience may need to support right-to-left (RTL) text, while a system intended for a global audience may need to accommodate different date formats (e.g., MM/DD/YYYY vs. DD/MM/YYYY).

Designing for Cultural Differences

Cultural differences can also influence users' expectations and preferences when it comes to software design. For example, users in some cultures may prefer formal, information-dense interfaces, while others may prefer minimalist, visually focused designs. Understanding these cultural nuances is essential for creating software that feels intuitive and comfortable for users around the world.

By incorporating cultural considerations into usability design, developers can ensure that their products are accessible, relevant, and appealing to a diverse global audience.

The Future of Usability Metrics

As usability practices continue to evolve, so too must the methods used to measure and assess usability. Traditional metrics like task completion time, error rates, and user satisfaction will remain important, but emerging technologies and interaction paradigms will require more sophisticated metrics that capture the full user experience.

Emotional and Cognitive Usability Metrics

In the future, usability metrics will increasingly focus on emotional and cognitive responses to software. Tools that analyze users' emotional states—such as facial expression recognition, voice analysis, or physiological signals—can provide deeper insights into how users feel while interacting with a system. These metrics can help designers understand not only how efficiently users complete tasks but also how they experience the interface on an emotional level.

By incorporating emotional and cognitive metrics into usability assessments, designers can create more engaging, satisfying experiences that go beyond basic functionality.

AI-Powered Usability Analytics

AI-powered usability tools are already being used to analyze user behavior and identify areas for improvement. These tools can track user interactions in real time, providing data on everything from mouse movements to click patterns and time spent on specific tasks.

Importantly, these AI-driven tools offer significant time savings for design teams by automating repetitive analysis tasks, such as identifying common user errors or bottlenecks in navigation flows. This allows designers to focus their attention on the more strategic and creative elements of usability, such as developing innovative interaction paradigms and crafting emotionally resonant experiences.

In the future, AI-driven usability analytics will become even more sophisticated, using machine learning algorithms to predict usability issues and recommend design improvements based on user behavior patterns. This will enable designers to iterate more quickly and make data-driven decisions that improve the overall user experience.

Challenges Facing the Future of Usability

Despite the advancements in usability practices and tools, several challenges remain. As digital systems become more complex and user expectations continue to rise, designers must navigate new obstacles to ensure that usability remains a top priority.

Managing Complexity in Feature-Rich Systems

As software systems become more feature-rich, maintaining simplicity and usability becomes increasingly difficult. **Feature creep**—the gradual addition of new features over time—can lead to bloated interfaces that confuse users and detract from the overall experience.

To address this challenge, usability designers must focus on maintaining a balance between functionality and simplicity. This may involve prioritizing core features, reducing unnecessary complexity, and ensuring that the interface remains intuitive and easy to navigate, even as new features are added.

Ensuring Usability Across Diverse User Groups

As digital products reach a global audience, ensuring that they are usable by diverse user groups with different cultural, linguistic, and accessibility needs becomes a significant challenge. Creating inclusive, adaptable interfaces that work well for all users will require careful planning, testing, and iteration.

Usability designers must account for differences in technology access, literacy levels, and cultural norms, ensuring that their products are accessible and relevant to a wide range of users.

The Role of Collaboration in Advancing Usability

The future of usability will depend on collaboration between designers, developers, researchers, and users. By working together, these groups can share insights, identify usability challenges, and develop innovative solutions that improve the user experience.

Cross-Disciplinary Collaboration

Usability is a multidisciplinary field that draws on knowledge from psychology, design, computer science, and more. Collaboration between these disciplines is essential for advancing usability practices and ensuring that digital products are designed with the user in mind.

For example, collaboration between cognitive psychologists and interaction designers can lead to a deeper understanding of how users process information and make decisions, which can inform the design of more intuitive and efficient interfaces.

User Involvement in Design

User involvement will continue to be a critical factor in improving usability. By involving users throughout the design and development process—through user research, usability testing, and feedback loops—designers can ensure that the final product meets the needs and expectations of its target audience.

Continuous Evolution: Usability as a Dynamic Discipline

Usability is not a static discipline—it is constantly evolving in response to new technologies, user behaviors, and design trends. As we move forward, usability will continue to be a dynamic field that requires continuous learning, adaptation, and innovation.

The Importance of Ongoing Learning

For usability practitioners, staying up-to-date with the latest advancements in technology and design is essential. Whether through professional development, collaboration with peers, or user research, ongoing learning will be key to keeping pace with the rapidly changing digital landscape.

As new technologies like AI, AR, VR, and multimodal interfaces continue to emerge, usability designers will need to develop new strategies and techniques to ensure that these systems remain intuitive and accessible. This may involve experimenting with new interaction patterns, testing new methods for gathering user feedback or exploring innovative ways to measure user satisfaction.

Usability and Business: ROI and Competitive Advantage

The business implications of usability are profound, as user-friendly systems not only increase customer satisfaction but also drive greater adoption and retention. Usability has a direct impact on return on investment (ROI) and can provide companies with a significant competitive advantage in crowded markets.

Usability as a Competitive Differentiator

In today's digital marketplace, where users have an abundance of options, usability is often the key differentiator between successful products and those that fail. Users are more likely to adopt, recommend, and remain loyal to products that offer seamless, intuitive experiences. This is especially true for consumer-facing software, where poor usability can lead to high abandonment rates and negative reviews.

Reducing Costs Through Usability

Investing in usability early in the development process can lead to significant cost savings in the long term. By identifying and addressing usability issues during the design phase, companies can avoid costly redesigns, reduce customer support inquiries, and minimize user churn. Good usability also leads to higher employee productivity and customer satisfaction, further enhancing business outcomes.

Beyond the Screen: Usability in Emerging Sectors

Usability is expanding beyond traditional software systems to play a critical role in emerging sectors such as healthcare, education, automotive, and the Internet of Things. As digital technologies become integrated into more aspects of daily life, usability will be essential for ensuring that these systems are safe, effective, and user-friendly.

Usability in Healthcare

In healthcare, usability is more than just a convenience—it can be a matter of life and death. Poorly designed interfaces in medical devices, electronic health records (EHR) systems, or telemedicine platforms can lead to errors, miscommunications, and delays in care. Ensuring that these systems are easy to use, provide clear feedback, and minimize the risk of errors is critical for patient safety and care quality.

Usability in Automotive Technology

As cars become more connected and autonomous, the usability of in-car systems is becoming increasingly important. In-vehicle infotainment (IVI) systems, driver-assistance features, and self-driving technologies all require intuitive interfaces that minimize driver distraction and enhance safety. Designing usable automotive interfaces will be key to ensuring a smooth transition to fully autonomous vehicles.

Usability in IoT and Smart Devices

The Internet of Things (IoT) is creating new challenges for usability as users interact with a growing number of connected devices in their homes, workplaces, and cities. Ensuring that these devices are easy to set up, control, and monitor will be essential for widespread adoption. IoT devices must also provide clear feedback, allowing users to understand the status of their systems and take action when needed.

Final Reflections on the Road Ahead

The road ahead for usability is filled with exciting possibilities. As technology continues to advance, usability will play an even more central role in shaping how we interact with the digital world. From AI-powered systems and multimodal interfaces to personalized experiences and inclusive design, the future of usability promises to be both challenging and rewarding.

By staying true to the core principles of usability—while embracing new technologies, user needs, and design trends—designers and developers can create software that not only functions efficiently but also resonates with users on a deeper, more meaningful level. The future of usability is bright, and its role in shaping the next generation of digital experiences cannot be overstated. The road ahead may be complex, but the possibilities for innovation, improvement, and human connection are limitless.

Index

A

Q

T

W, X, Y